RESIDENTIAL ELECTRICAL WIRING

CLYDE N. HERRICK
SAN JOSE CITY COLLEGE

A Goodyear Applied Technical Book
GOODYEAR PUBLISHING COMPANY, INC.

Library of Congress Cataloging in Publication Data

Herrick, Clyde N
  Residential electrical wiring.

  1. Electric wiring, Interior. I. Title.
TK3285.H47 621.319'24 75-26468
ISBN 0-87620-800-6
ISBN 0-87620-799-9 pbk.

Current printing (last digit): 10 9 8 7 6 5 4 3 2 1

Y-8006-2
Y-7999-9 pbk.

Printed in the United States of America

Bro-Dart   2-24-77   15.00

CONTENTS

PREFACE

This book has been written to fill a growing demand for an electrical book that is completely practical and sufficiently comprehensive to answer all of the basic questions concerning residential wiring installations. The format is profusely illustrated and highly specific, but without theoretical discussions or undue length. It is assumed that the reader knows the fundamentals of electricity and how to work with ordinary tools. Considerable attention is given to the importance of planning. The topical development is functional, proceeding from planning to layout, material listing, service-entrance requirements, wiring installation, circuiting, and troubleshooting.

Materials are fully illustrated, installation drawings and photographs are provided, and circuit diagrams are presented where useful. Technical words and phrases are carefully explained. Great care has been taken to ensure that the treatment is accurate and up to date. Aside from a basic knowledge of electricity, the reader needs no background in electrical wiring practice. Mathematics has been almost completely avoided, inasmuch as a wireman has little practical use for calculations. Tables are provided where necessary for locating any data that would otherwise require calculation.

This book will find valuable use as a self-instructional guide, and also as a textbook in technical institutes. The electrical apprentice will find it highly informative, and the ambitious do-it-yourselfer can follow its instructions profitably. The beginner should read the book through before starting out on his first project. This initial "once over" will serve to bring the complete wiring system into focus and enable the reader to proceed with greater confidence. The beginner is also strongly advised to obtain a copy of the latest National Electrical Code for reference.

In addition to its self-instructional format, this book has been designed as a teaching tool for classroom instruction in junior colleges, high schools, technical institutes, and vocational schools. As noted, the text assumes only minimal prerequisites. A student who has completed the junior-college requirement in basic electricity is qualified for enrollment in an electrical-wiring course structured around this textbook.

This work is the outcome of extensive teaching experience, on the part of both the author and his fellow instructors at San Jose City College, who have contributed numerous constructive criticisms and suggestions. It is appropriate that this text be dedicated as a teaching tool to the instructors and students of our high schools, vocational schools, technical institutes, and junior colleges.

Clyde N. Herrick

PRINCIPLES OF ELECTRICITY 1

GENERATION AND DISTRIBUTION OF ELECTRICITY

Supplies of electricity first became available to the public in the 1880s, and the rate of growth has amounted to a doubling of the load every ten years into the 1970s. Primary energy sources are water power, coal, oil, natural gas, and nuclear energy. Large power stations (generating stations) operate at voltages as high as 700,000 volts AC and 1,500,000 volts DC. Thomas Edison's first distribution system, a low-voltage DC arrangement, was placed in operation in 1882. By 1898 alternating-current transmission had been developed to the extent that a 30,000-volt, seventy-five-mile line was placed in service from Los Angeles to Santa Ana. In 1899 a 40,000-volt, seventy-mile line from Sacramento to Colgate Hydro was opened. By 1920 lines operating at voltages up to 132,000 volts were common, and by 1945 220,000-volt operation came into use. In 1934 electric power generated at Boulder Dam in Colorado was transmitted to Los Angeles, a distance of 270 miles, with a line operating at 287,000 volts. Line efficiency and operating economy increased accordingly.

Although there is a practical limit to the voltage at which a power line can operate, studies and experiments on lines operating at 500,000 volts were made following World War II. By 1965 Russia had constructed 8,000 kilometers of 500,000-volt AC lines, followed by a 750,000-volt system. In the United States, a 765,000-volt system began operation. A 1.5-million-volt line is also being constructed in this country. In the late 1960s, there were approximately 300 privately owned electric power utilities in the United States, serving up to 3 million customers. The United States generating capacity at that time was about 320 million kilowatts, with a production of approximately 486 billion kilowatt-hours annually. By the end of the 1960s, electric utilities were generating over 1.4 trillion kilowatt-hours annually.

A nuclear reactor, or atomic reactor, is an arrangement for maintaining a controlled, self-sustaining nuclear chain reaction from fissionable or fusionable material. Fusion reaction occurs in the stars, but scientists have not been able to duplicate other than a very small version of stellar fusion reactor processes. Fusion reactor research is a very active area at this time. Fission reactors, however, have already made much greater progress. By the late 1960s, the number of fission reactors in the world exceeded 400. Constructed for a wide variety of uses, fission reactors vary in core size from a few inches to dozens of feet, and in power capability from a few watts to hundreds of megawatts (millions of watts). All reactors develop heat, neutrons, gamma rays, and fission products, but most applications utilize only one of these products. Electric generating stations employ heat from nuclear reactors to form steam, which in turn drives turbines for turning electric generators.

We should note that the principal release of energy in a reactor begins with fission fragments (subatomic particles), which fly apart and develop heat at a temperature of millions of degrees. In 1951 the Argonne National Laboratory constructed the first reactor to produce electric power on a commercial scale. In 1956 the first large reactor power plant was placed in operation by the United Kingdom at Calder Hall. These early generating stations lacked efficiency, but further development soon overcame this disadvantage. By 1969 the United States had 12 reactor power plants in operation, with 65 more in the construction or planning stage. This would bring the total nuclear capacity to

65 million kilowatts—equal to one-fourth of the national demand for electricity.

During the 1960s there was an emergence of increased public awareness of the environmental impact accompanying ever-growing sizes and numbers of transmission lines. Aesthetics began to make an important entry into the electrical-engineering field. Technical requirements for adequate transmission systems were restructured in the context of aesthetic effects on the environment. Relevance of transmission systems to social values in addition to traditional economic values led to a trend of installing transmission lines, distribution lines, and related facilities underground. A more immediate trend was the construction of aesthetically pleasing structures. The cost of installing transmission circuits underground is very high, and AC operation reduces efficiency. Therefore, intensive development work has been undertaken in experimental DC operation. A typical power system for a large city is shown on the inside cover of the text.

As an illustration, an experimental DC submarine cable transmission line operating at 100,000 volts and carrying 20,000 kilowatts of power was put into service between Sweden and Gotland, a distance of 60 miles, during 1954. In 1961, two 30-mile submarine cables were laid under the English Channel from England to France, operating at 200,000 volts DC with a capacity of 160,000 kilowatts. In the later 1960s, a 500,000-volt DC line had been installed in New Zealand, and in 1970 a 750,000-volt DC line 800 miles long with a capacity of 13.5 million kilowatts was installed in the western United States. At this time, Russia had constructed an 800,000-volt DC line 500 kilometers in length with a capacity of 720,000 kilowatts. It can be forecast that although high-voltage DC transmission lines will play a greater role in future electric-power transmission systems, high-voltage AC lines will retain a leading role.

GENERAL CONSIDERATIONS

An electrician must understand the principles of electricity in order to install electrical wiring correctly.

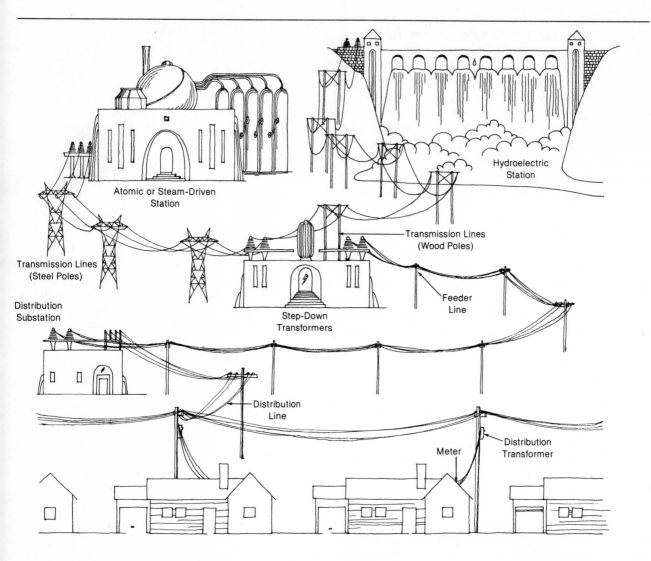

FIGURE 1-1   Example of generation, transmission, and distribution of electrical power

Although a wiring system seems mysterious and confusing to most people, it appears very simple to an experienced electrician. This apparent simplicity develops from an understanding of voltage, current, resistance, power, and electric circuits. Therefore, it is necessary for us to review the basic principles of electricity at this point. When we speak of electricity, we often mention magnetism also, for the two are very closely related. Sometimes a student asks, "What is electricity?" or "What is magnetism?" These questions are still unanswered in some respects. For example, we can say that electricity is the force that moves electrons, but this is the same as saying that an engine is the force that moves an automobile. We have described the effects of forces, but we failed to describe the forces themselves. In spite of this difficulty, if we learn the rules or laws that apply to the behavior of electricity and magnetism, we will have "learned" electricity without ever having determined its fundamental identity.

Most people are more or less familiar with magnets. Natural magnets are called lodestones and have the ability to attract iron objects, as ilustrated in Figure 1-2. Lodestones were known to the ancients. Although early compasses were made from lodestones, modern compasses (Figure 1-3) use steel needles that have been magnetized. Lodestones and compass needles are examples of permanent magnets; this means that the magnet maintains its magnetism for an indefinite length of time. We can make a steel needle into a permanent magnet by means of electricity, as will be explained later.

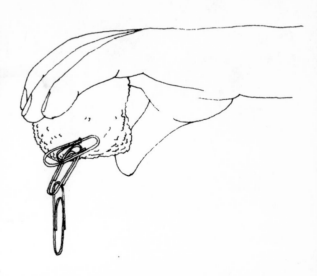

FIGURE 1-2   A lodestone has magnetic attraction for iron objects

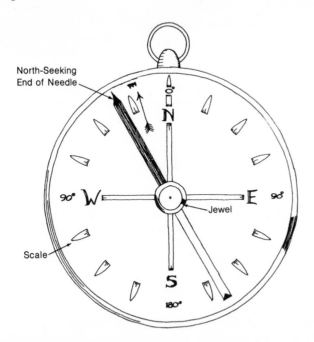

FIGURE 1-3   Typical magnetic compass

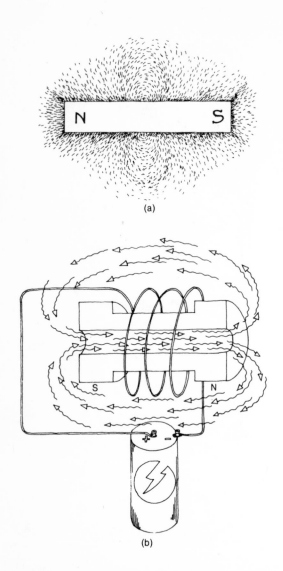

(a)

(b)

FIGURE 1-4   Magnetic field representation by lines of force
(a) Lines of force around a permanent magnet   (b) Lines of
force around an electromagnet

Although magnetic forces are invisible, it is helpful to indicate a magnetic field by drawing lines of force, as shown in Figure 1-4(a). Lines of force can also be demonstrated physically. For example, if we place a sheet of cardboard over a magnet and sprinkle iron filings on the cardboard, the filings will arrange themselves as depicted in Figure 1-4(a). It is usually necessary to tap the cardboard lightly, so that the filings vibrate into a completed pattern. Next, observe the electromagnet arrangement shown in Figure 1-4(b). Several turns of wire are wound around a wooden spool. When we connect this coil to a battery, we produce a magnetic field, as indicated by the lines of force. However, this is not a permanent magnet; it is a temporary magnet, or electromagnet. When we disconnect the coil from the battery, the magnetic field no longer exists.

Note that the electromagnet depicted in Figure 1-4(b) is an important example of the relationship between electricity and magnetism. That is, when an electric current flows (electrons move) around the coil winding, it produces magnetism. In other words, magnetism is electricity in motion. When the flow of electricity stops, neither the coil nor the battery has any magnetic property. Beginning students often ask, "If magnetism is electricity in motion, how can a permanent magnet produce magnetism?" The answer to this good question is that, in part, the atoms of steel consist of orbiting electrons. In turn, these moving electrons represent small electric currents, and these currents produce magnetism. Some metals such as copper are not magnetic because the magnetic fields produced by orbiting copper atoms are directed so that they cancel out on the average.

An electromagnet with a core of insulating material or of air is often called a *solenoid*. Dishwashers and various other electrical equipment use solenoids. When an iron core is placed in a solenoid, it has the effect of increasing the field strength of the electromagnet considerably. Figure 1-5(a) shows a simple example. The electromagnet can suspend quite a few more tacks against the force of gravity when an iron spike is placed inside the coil of wire. Each end of the electromagnet will suspend the same number of tacks. However, the two ends of an electromagnet are not exactly the same, because one end is a north pole and the other end is a south pole. A compass shows this fact, as pictured in Figure 1-5(b). In other words, the south pole of the electromagnet will attract the north-seeking end of the compass. On the other hand, the north pole of the electromagnet will repel the north-seeking pole of the compass. *Unlike poles attract, and like poles repel.*

Next, suppose that we replace the iron spike in Figure 1-5(a) by a steel rod. As before, the end of the steel rod will attract tacks when electricity flows through the coil. However, we must note an important fact: when we stop the current flow through the coil, the tacks remain suspended from the end of the steel rod. In other words, we have formed a permanent magnet. This is one of the basic ways of making permanent magnets. To produce a very strong permanent magnet a large number of turns are required on the coil; each turn on the coil adds strength to the magnetic field. We will find in the next section that a stronger battery is also required to force a larger current through a coil.

BASIC ELECTRIC CIRCUITS

A simple example of an electric circuit is shown in Figure 1-5. An electric circuit is a complete (closed) conducting path for electron flow. It not only consists of the conductors (connecting wires), but it also includes the path through the voltage source (battery). Electricians often say that electric current flows through the battery from the positive terminal and flows out of the negative terminal. Electrons will not flow in this manner, however, unless the positive and negative terminals of the battery are connected to a closed circuit. Figure 1-6 illustrates this requirement. Note that current flows from the (−) terminal of the battery through the lamp to the (+) battery terminal, and continues by going through the battery from the (+) to the (−) terminal. As long as this pathway is unbroken, it is a closed circuit and current will flow. On the other hand, if the path is broken at *any* point, it is an open circuit, and no current flows.

Let us add a practical note: current will not flow in the foregoing closed circuit if the battery is "dead." In

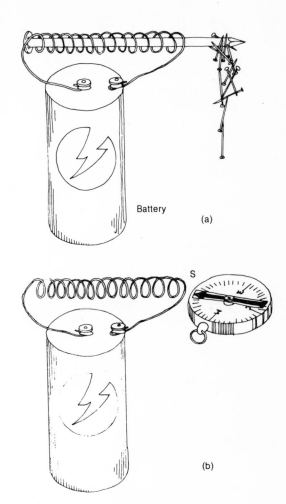

Battery

(a)

S

(b)

FIGURE 1-5 Properties of electromagnets (a) An iron core increases the strength of an electromagnet (b) A compass needle indicates the poles of an electromagnet

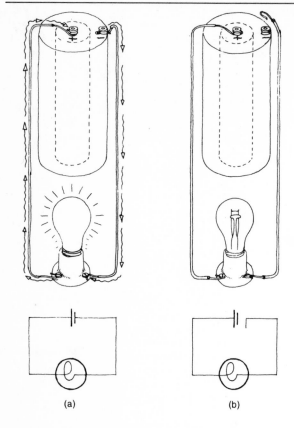

(a)   (b)

FIGURE 1-6   Basic electric circuits   (a) Closed circuit
(b) Open circuit

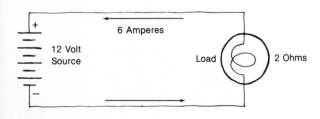

FIGURE 1-7   The lamp is the load in this simple circuit

other words, the battery uses up chemical energy to move electrons from its positive terminal to its negative terminal; as electrons move through the battery, this chemical energy will eventually be reduced to zero. We can see from the facts already discussed that a closed loop of wire (conductor) is not necessarily a circuit. That is, we must include a source of voltage (electrical pressure) to provide an electric circuit. Then, if the circuit is closed, current will flow because this voltage overcomes the opposition (resistance) of the wire to electron flow. It follows that the amount of current (amperes) that flows in the wire will be greater if we increase the voltage (volts) applied to the circuit. Or the amount of current flow will be greater if we decrease the resistance (ohms) of the circuit.

If we know the voltage and resistance of a circuit, we can calculate the resulting current flow. If we know the resistance and current flow in a circuit, we can calculate the applied voltage. Or if we know the current flow and voltage of a circuit, we can calculate its resistance. These basic relations are stated in the form of Ohm's law: **The intensity of the current in amperes in any electric circuit is equal to the difference in potential in volts across the circuit divided by the resistance in ohms of the circuit.** Ohm's law is fundamentally expressed as a simple equation:

$$\text{Current} = \frac{\text{Voltage}}{\text{Resistance}} = \text{Amperes} = \frac{\text{Volts}}{\text{Ohms}} = I = \frac{E}{R}$$

Thus, I is the intensity of the current in amperes, E is the difference in potential in volts, and R is the resistance in ohms. One ampere of current flows when 1 volt of potential difference is applied across 1 ohm of resistance. As Figure 1-7 shows, if the voltage across the lamp is 12 volts, and the effective resistance of the lamp is 2 ohms, the current flow will be 12/2, or 6 amperes. Note also that if we reduce the applied voltage to zero, the current flow will be equal to 0/2, or 0 amperes. It is evident that if the applied voltage is 12 volts and the current flow is 6 amperes, the resistance is then equal to 2 ohms, by Ohm's law:

$$R = \frac{E}{I} = \frac{12}{6} = 2 \text{ ohms}$$

Or, if the current flow is 6 amperes, and the resistance is 2 ohms, the applied voltage is then equal to 12 volts, by Ohm's law:

$$E = IR = 6 \times 2 = 12 \text{ ohms}$$

The Ohm's law diagram in Figure 1-8 depicts an easy way to remember the foregoing facts. If you place your finger over the item you want, the remaining part of the diagram shows the formula for obtaining that item.

ELECTRIC POWER AND ENERGY

Power is the rate at which work is being done. For example, if a force of 1 pound is exerted on an object, and this force lifts the object through one foot in 1 second, 1 foot-pound of work is being done in 1 second. In other words, the power expended is equal to 1 foot-pound/sec. Note that 1 horsepower (hp) is equal to 550 foot-pounds/sec. This is an important unit of power for the electrician, because electric motors are rated in horsepower.

Now let us consider electrical power. We know that voltage is an electrical force and that voltage forces current to flow in a closed circuit. Note that unless current flows, no work is done—no electrical power is expended. However, when current flows, electrical work is being done, and this electrical work rate is measured in watts. If one volt is applied to a load and 1 ampere flows, the electrical work rate, or power, is equal to 1 watt. Note also that 1 horsepower is equal to 746 watts.

Beginning students sometimes tend to confuse power with work. However, the distinction becomes clear if we recognize that an electric lamp might expend a power of 1 watt for a second, a minute, an hour, a day, or longer. If the lamp is turned on for 1 second, the amount of work accomplished is 1 watt-second. Or if the lamp is turned on for a minute, the work accomplished is 60 watt-seconds. Again, if the lamp is turned on for 1 hour, the work accomplished is equal to 1 watt-hour. Work is also called energy, and 1 watt-hour of work is the same thing as 1 watt-hour of

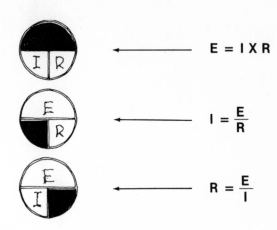

$$E = I \times R$$

$$I = \frac{E}{R}$$

$$R = \frac{E}{I}$$

FIGURE 1-8   The three forms of Ohm's law

energy. Most of us are familiar with watt-hour meters, such as the one shown in Figure 1-9. A watt-hour meter indicates the amount of electrical energy that has been consumed. It is customary to state this energy in terms of kilowatt-hours (kwh). A kilowatt is equal to 1,000 watts.

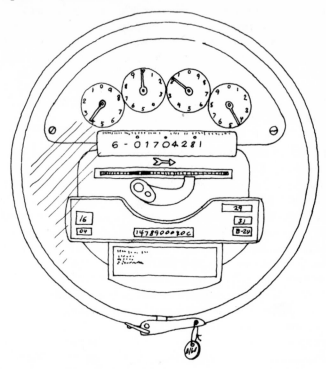

FIGURE 1-9   Kilowatt-hour meter

Electrical lamps, soldering irons, and ranges are examples of electrical devices that are rated in watts. This wattage rating indicates the rate at which the device converts electrical energy into another form of energy, such as heat or light. As an illustration, a 100-watt lamp produces a brighter light than a 75-watt lamp, because it converts more electrical energy into light. Electrical energy is converted into heat when current flows through resistance. If the heat is sufficiently intense, light will also be produced. Figure 1-10 shows that a lamp has a resistive filament, an electric iron has a heating coil, an electric range has a resistance element, and water heaters also have resistance

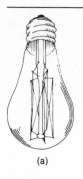
(a)

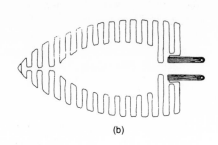
(b)

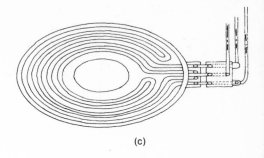
(c)

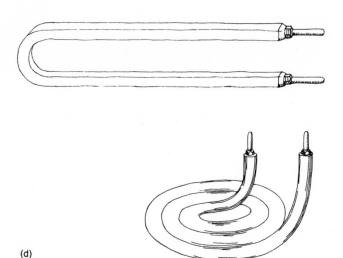
(d)

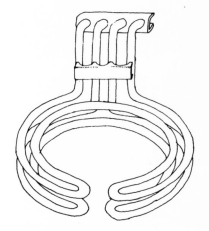

elements. It is important for an electrician to know the relationship of voltage, current, resistance, and power.

We have said that electrical power in watts is equal to the product of the voltage in volts times the current in amperes:

FIGURE 1-10   Typical resistive electrical devices   (a) Incandescent lamp   (b) Electric flatiron   (c) Resistance element for electric range   (d) Resistance elements for water heaters

$$\text{Power} = P = EI \text{ watts}$$

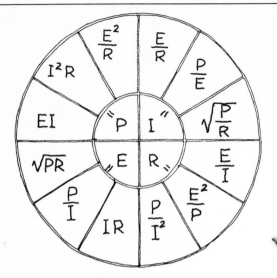

FIGURE 1-11   Diagram for remembering the power laws

If we substitute the terms of Ohm's law into the foregoing power formula, we obtain the following equivalent expressions:

$$P = I^2R \text{ watts}$$

$$P = \frac{E^2}{R} \text{ watts}$$

Figure 1-11 provides a summary of these power equations and their rearrangements. An electrician should know these twelve basic expressions, in addition to the three forms of Ohm's law. The diagram helps in remembering the equations; however, students are advised not to neglect using the algebra that they have learned merely because a memory aid is available. In practical work, situations and problems will sometimes arise in which the electrician would be lost if he had forgotten his basic algebra.

ELECTRIC CIRCUITS AND DEVICES

We can show electrical circuits either in pictorial form or in schematic form. As an illustration, Figure 1-12 depicts both schematic and pictorial diagrams for a

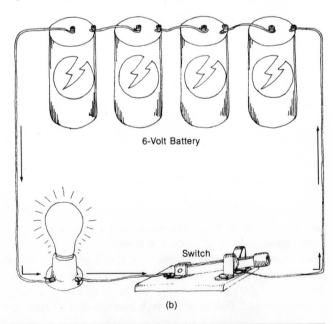

6-Volt Battery

Switch

(b)

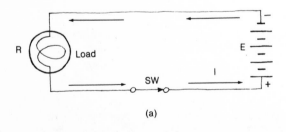

(a)

FIGURE 1-12   Two types of electrical diagrams   (a) Pictorial diagram   (b) Schematic diagram

simple circuit comprised of a lamp, a battery, and a switch. Observe that the devices are shown as symbols in (a), but are depicted as objects in (b). Electricians generally work from schematic diagrams, because a schematic diagram can be prepared in a small fraction of the time that is required to prepare a pictorial diagram. After a person learns how to read schematic diagrams, they provide the same information as the pictorial diagrams.

Note that the 6-volt battery in Figure 1-12 consists of four cells. Each cell supplies 1.5 volts. Since the cells are connected in "series-aiding" (positive terminal of one cell to the negative terminal of the next cell), their total voltage is 6 volts. A cell is often called a battery, although, in the strict sense of the term, a battery consists of two or more cells. The switch depicted in Figure 1-12 is called a knife switch. Since the battery supplies 6 volts, the lamp must be rated for 6-volt operation. If a 3-volt lamp were used, it would promptly burn out. Or if a 12-volt lamp were used, it would glow very dimly. Note that the lamp is mounted in a socket. Electrons flow around the circuit in the direction of the arrows.

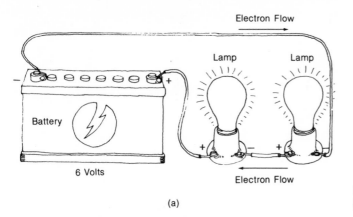

(a)

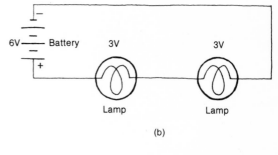

(b)

FIGURE 1-13   Load consisting of series-connected lamps
(a) Pictorial diagram   (b) Schematic diagram

Figure 1-13 shows another pictorial diagram and its corresponding schematic diagram. In this example, the load consists of two lamps connected in series. Since the battery applies 6 volts to the series-connected lamps, each lamp must be rated for 3-volt operation. In other words, half of the battery voltage "drops" across each of the lamps. Series-connected lamps are used in strings of Christmas tree lights, for example. Figure 1-14 illustrates a string of eight lamps. Note that the circuit is energized from a 120-volt source, and that each lamp is rated for operation at 15 volts.

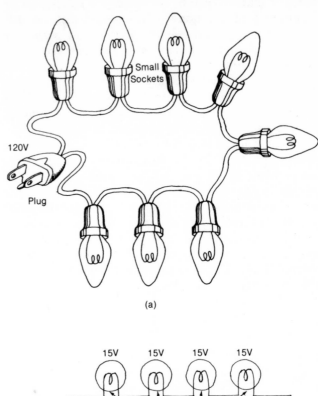

FIGURE 1-14   Circuit comprised of eight series-connected lamps   (a) Pictorial diagram   (b) Schematic diagram

Let us now consider the effects of accidental short-circuits in the doorbell arrangement of Figure 1-15.

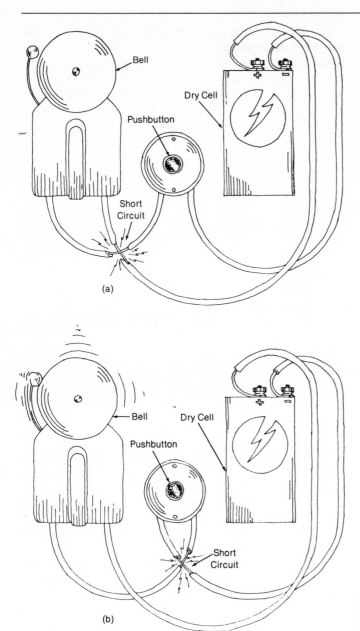

FIGURE 1-15   Examples of short-circuits   (a) Doorbell terminals are short-circuited   (b) Pushbutton switch terminals are short-circuited

In (a), defective insulation on the wires has caused a short-circuit across the bell terminals. Hence, the bell will not ring when the pushbutton switch is depressed.

Next, in (b), defective insulation on the wires has caused a short-circuit across the pushbutton switch terminals. Consequently, the bell will ring continuously, although the pushbutton switch is not depressed. Note that another type of short-circuit could occur: defective insulation on the wires could cause a short-circuit across the battery terminals. In such a case, the bell would not ring when the pushbutton switch was depressed. Furthermore, the battery would soon "go dead" because of the large current that would flow through the short-circuit.

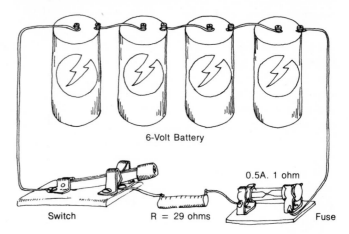

FIGURE 1-16   A fuse is connected in series with the circuit

Short-circuits can cause fires by overheating the wires that carry the short-circuit current. Therefore, most wiring systems contain protective devices such as fuses. A fuse will blow in case of a short-circuit, automatically opening the circuit. Thus, we can regard a fuse as an automatic switch. Figure 1-16 shows a simple circuit that includes a fuse. In normal operation, the fuse will not blow. However, in case the 29-ohm load resistor happens to become short-circuited, the fuse will blow. (Note that omega, Ω, is a symbol for ohms). Observe that the total circuit resistance consists of the load resistance, the fuse resistance, the connecting-wire resistance, the internal resistance of the battery, and the switch resistance. However, from a practical viewpoint, we can regard the total circuit resistance as the load resistance plus the fuse resistance, or 30 ohms. In normal operation, the current flow will be:

$$I = \frac{6}{30} = 0.2 \text{ ampere}$$

Next, suppose that the load resistor becomes short-circuited. Then the current flow will be:

$$I = \frac{6}{1} = 6 \text{ amperes}$$

Since the fuse is rated for 0.5 ampere, it will not blow with 0.2 ampere circuit current flowing. On the other hand, a circuit current of 6 amperes is 30 times the fuse rating, and it will blow instantaneously. Referring to Figure 1-15, note that it is advisable to install a fuse near the voltage source (battery) in order to minimize the possibility of an unfused short-circuit condition. As an illustration, suppose we installed the fuse between the switch and the load in Figure 1-16. Then, if a short-circuit occurred between the wires connected to the positive terminal and to the negative terminal of the battery, the fuse would not blow. In other words, a fuse provides greatest protection when it is installed as near the voltage source as possible. Figure 1-17 shows some basic types of fuses used in electrical wiring systems.

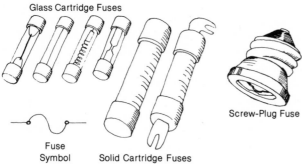

Glass Cartridge Fuses

Fuse Symbol   Solid Cartridge Fuses

Screw-Plug Fuse

FIGURE 1-17   Some basic types of fuses

Next, carefully observe the *parallel circuit* depicted in Figure 1-18. It consists of two branch circuits and three lamps, each with its on-off switch. Note that branch circuit (2) is connected in parallel with branch circuit (1). Each lamp is connected across the line; the lamps are *not* connected in series. This type of circuitry is used in practically all residential wiring systems.

Switch No. 1 is a main switch; if it is opened, none of the lamps can be energized. In normal operation, switch No. 1 is closed, and the lamps are turned on or off as desired by throwing switch No. 2, 3, or 4. The chief advantage of a parallel circuit is that if one lamp burns out, the other lamps will continue to glow. On the other hand, if one lamp burns out in a series circuit, all the lamps stop glowing until the defective lamp is replaced. Moreover, individual lamps cannot be operated in a series circuit: either all lamps are turned on, or all are turned off.

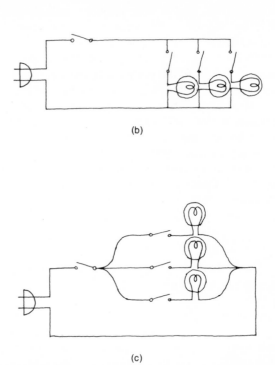

(b)

(c)

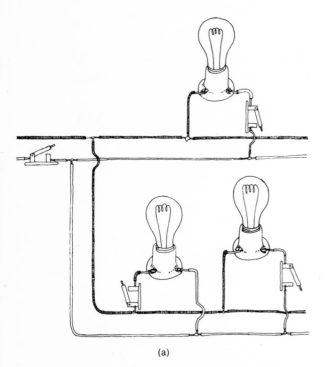

(a)

FIGURE 1-18   Basic parallel circuit   (a) Pictorial diagram
(b) Simplified equivalent circuit(c) Another equivalent
circuit

Now consider the operation of the parallel circuit shown in Figure 1-19. Five lamps are energized from a 120-volt source. Each lamp operates at 120 volts and draws 1 ampere of current. Thus, the total current flow is 5 amperes. Each lamp consumes 120 watts, and the total lamp load is 600 watts. Note that voltage V1 is equal to zero (assuming that the wire has negligible resistance). On the other hand, voltage V2 is 120 volts. Again, voltage V3 is equal to zero. However,

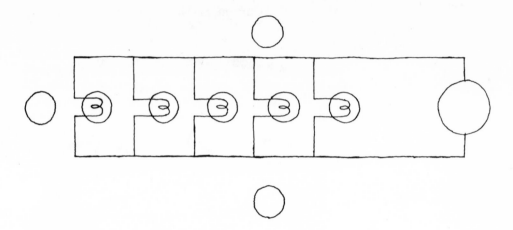

FIGURE 1-19 Current flow and voltages in a parallel circuit

suppose that the connecting wires are long and have a small diameter. In such a case, the line will have appreciable resistance. The practical results are, first, that lamp No. 5 will glow dimly compared to lamp No. 1. Second, the connecting wires will tend to run hot, owing to the I^2R loss in their resistance. There will be a voltage drop along the line; in other words, voltmeters V1 and V2 might indicate a loss of 10 volts along the run from the generator to the last lamp. The way to avoid this kind of trouble is to install wires that have sufficiently large diameter. We shall discuss this topic in greater detail later.

ELECTRICAL METERS

Electricians use electrical meters at various times, particularly when it is necessry to locate a fault in a wiring system. It is very important to use electrical meters correctly, because a meter can be damaged if it is applied improperly. The most important instrument is a voltmeter. Figure 1-20 illustrates a typical volt-ohm-milliammeter of the kind used by wiremen, appliance technicians, and electrical repairmen. A milliammeter is used to measure small values of current. A volt-ohm-milliammeter (VOM) measures volts, ohms, and milliamperes or amperes. Note that a milliampere is one one-thousandth of an ampere. A wireman usually checks the voltage in a circuit. However, he may also need to check the resistance of a circuit or of parts of a circuit. Current measurements are seldom made by wiremen.

FIGURE 1-20   Volt-ohm-milliammeter used by various kinds of electricians

It follows from previous discussion that a voltmeter is always connected *across* the circuit or device, as shown in Figure 1-21(a). On the other hand, an ammeter (used to measure circuit current) is always connected *in series* with the circuit or device, as depicted in Figure 1-21(b). Remember that if a voltmeter were connected in series with a circuit or device, such as a lamp, the lamp would not glow and the circuit would seem to be defective. This results from the fact that a voltmeter has a very high internal resistance. Remember also that if an ammeter were connected across a circuit or device, the meter would immediately burn out. This would occur because an ammeter has a very low internal resistance. The final rule to remember is that an ohmmeter must always be applied in a *dead circuit*. In other words, if the circuit is "live," the voltage in the circuit will immediately burn out an ohmmeter. Applications of meters and other indicating devices are explained in greater detail in a later chapter.

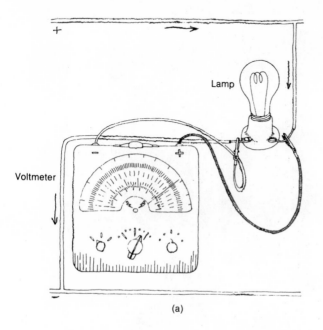

(a)

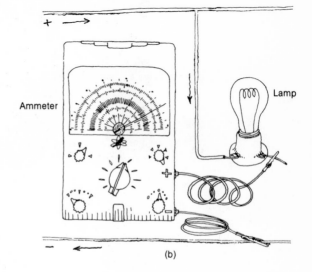

(b)

FIGURE 1-21   Application of voltage and current meters   (a) A voltmeter must be connected across the lamp   (b) An ammeter must be connected in series with the lamp

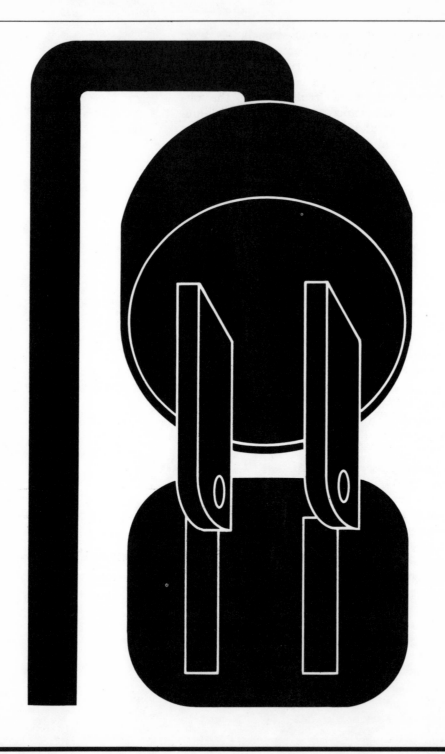

FUNDAMENTALS OF ALTERNATING CURRENT 2

GENERAL CONSIDERATIONS

In our previous discussion of electric circuits we have been concerned with direct-current (DC) operation. It is helpful to start with consideration of DC circuits, because they are comparatively simple. On the other hand, all of the circuits with which a wireman is concerned are alternating-current (AC) circuits. Since there is a close relationship between AC and DC circuit operation, all of the facts and laws that we learned in the first chapter will help us to understand the operation of AC circuits. The chief difference between direct current and alternating current is that AC does not have a fixed polarity. In other words, an AC voltage is first positive, then negative, and then positive again. Thus, the AC voltage swings back and forth, or *alternates* in polarity, as depicted in Figure 2-1. We will find that if alternating current is considered at any particular instant, it can be regarded as direct current.

Figure 2-1 shows that the AC voltage rises first to a maximum or peak value of about 150 volts positive, and then falls back to zero. Then, the AC voltage reaches a peak value of about 150 volts negative, after which it again falls back to zero. Most AC lines operate at a frequency of 60 Hertz (Hz). This means that the changes shown in Figure 2-1 occur in 1/60 second; that is, the AC voltage goes through a positive peak 60 times a second.

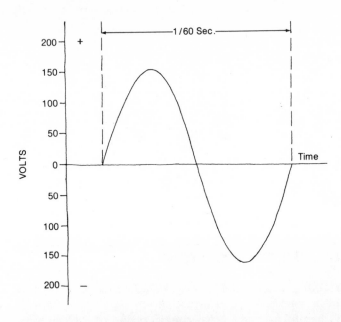

FIGURE 2-1   Representation of AC voltage

Beginning students usually ask why AC is preferred instead of DC for operation of residential wiring systems. The answer is that AC can be used to operate *transformers*, whereas DC cannot. Figure 2-2 shows examples of a line transformer and a bell-ringing transformer. A line transformer might reduce the voltage at the crossarm from 1200 volts to 120 volts for energizing a residential wiring system. Again, a bell-ringing transformer reduces 120 volts to about 8 volts for operating door bells or chimes.

Note the general construction of a transformer, as seen in Figure 2-3. A simple transformer has two windings on an iron core. These windings are called the primary and the secondary. If there are more turns on the primary than on the secondary, we call the transformer a step-down transformer. On the other hand, if there are more turns on the secondary than on the primary, we call the transformer a step-up transformer. This means that the secondary voltage is less than the

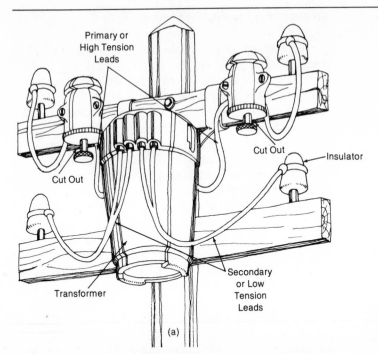

Primary or
High Tension
Leads

Cut Out

Cut Out

Insulator

Transformer

Secondary
or Low
Tension
Leads

(a)

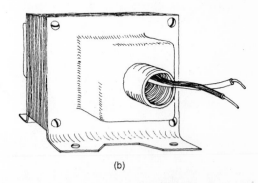

(b)

FIGURE 2-2   Examples of transformers   (a) Line transformer
(b) Bell-ringing transformer

primary voltage in the case of a step-down transformer.
On the other hand, the secondary voltage is greater
than the primary voltage in the case of a step-up
transformer. Both of the examples depicted in Figure
2-2 are step-down transformers. It is interesting to
note that the voltage step-down (or step-up) is the
same as the ratio of turns on the coils. For example, if
the primary has 500 turns, and the secondary has 50
turns, the secondary voltage will be 1/10 of the primary voltage. Remember this important fact: a transformer will operate on AC, but it will not operate on
DC.

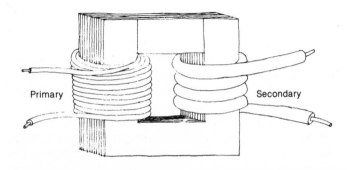

Primary

Secondary

FIGURE 2-3   Arrangement of a simple transformer

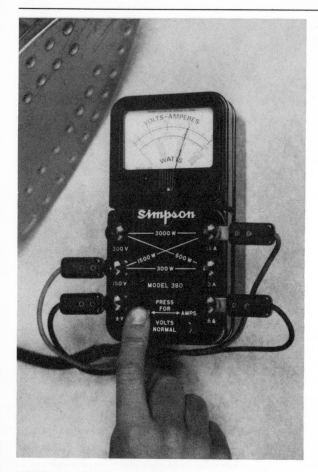

FIGURE 2-4   Volt-amp-wattmeter for use in AC circuits

FIGURE 2-5   Neon-light tester used by wiremen for checking circuits and fuses

AC METERS AND INDICATORS

We mentioned the volt-ohm-milliammeter (VOM) in the preceding chapter. A VOM has both DC and AC functions; we cannot measure AC voltage on the DC function of the instrument. In other words, if we connect a DC voltmeter to an AC line, we obtain no indication. When a VOM is operated on its AC-voltage function, it will indicate the value of AC voltage that is applied. Note that an AC-voltage reading can be used in Ohm's law, just as a DC-voltage reading can. Thus, if we apply 120 volts AC to a lamp, it will glow just as brightly as if 120 volts DC were applied. We can also calculate power values on the basis of AC-voltage measurements. Although a VOM cannot measure AC current values, electricians often use meters such as the one illustrated in Figure 2-4 to measure how much AC current is being drawn by a circuit or a load.

The instrument shown in Figure 2-4 is a volt-amp-wattmeter; that is, it can measure AC voltage, AC current, and AC power. For example, suppose that an electrician wishes to check the load that is imposed by a kitchen appliance. He unplugs the appliance and plugs the volt-amp-wattmeter into the electrical output. In turn, he plugs the appliance into the instrument receptacle. The meter then reads the voltage that is being applied. Next, if the electrician presses the current button on the instrument, the meter indicates the amount of current being drawn by the appliance. Then, if he presses the power button on the instrument, the meter indicates the power in watts that is being consumed by the appliance. If the appliance is turned off and on, the line voltage will be indicated under no-load and full-load. An efficient wiring installation will have practically the same line voltage at full load as at no load.

When accurate measurements are not required, and the electrician wishes only to check for the presence or absence of AC voltage, he often uses a neon indicator such as that illustrated in Figure 2-5. It will glow if it is applied across a circuit that has 65 volts or more present. A neon indicator is convenient because it is small and rugged, in comparison to a meter. Another indicator that an electrician often uses is shown in Figure 2-6. This is an ordinary doorbell that has been taped to a battery and provided with a pair of test leads. If the leads are connected together, the bell will

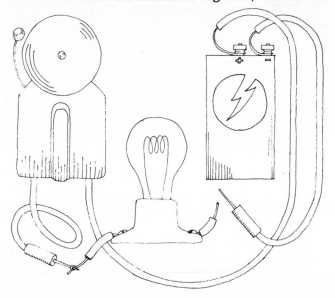

FIGURE 2-6 Dry-cell and doorbell arrangement for continuity testing

ring. A wireman uses this arrangement to check for continuity of wire installations. For example, if a pair of wires is run from the front to the back of a residence, various connections may be made along the run, as we shall explain later. To check the installation, the far ends of the wires may temporarily be short-circuited. Then, if the bell rings when the near ends of the wires are tested, the electrician knows that circuit continuity is present. Note that an ohmmeter can also be used to check continuity.

THREE-WIRE AC LINES

Most residential wiring systems use three-wire circuits. This type of circuiting is preferred because it provides a choice of either 120-volt or 240-volt operation and reduces wire cost. To understand the principle of a three-wire circuit, examine the circuits shown in

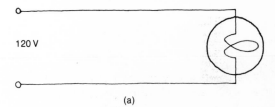

120 V

(a)

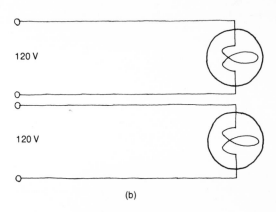

120 V

120 V

(b)

FIGURE 2-7 Basic circuit arrangements (a) Single 120-volt circuit (b) Pair of separate 120-volt circuits

Figure 2-7. Two wires are employed in the circuit shown at (a). Next, four wires are utilized in the pair of circuits shown at (b). This pair of circuits could be energized from separate sources, or they could be energized from a line transformer with a center-tapped secondary, as depicted in Figure 2-8. This secondary provides 240 volts across its entire winding, or 120 volts from either end to the center tap. At a particular instant, the electrons flow in this circuit as indicated by the arrows. Notice that the arrows along the wires to the center tap point in opposite directions. Since these currents will cancel out, we can replace this pair of wires to the center tap by one wire, and save the cost of an additional wire.

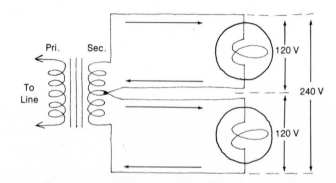

FIGURE 2-8   Pair of circuits energized from a center-tapped transformer

Figure 2-9 shows the foregoing pair of circuits re-arranged as a three-wire system. This system has a

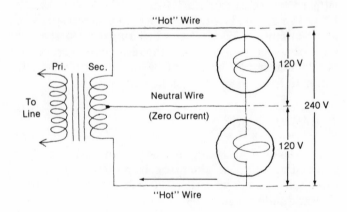

FIGURE 2-9   A three-wire circuit has two "hot" wires and a neutral wire

neutral wire and a pair of "hot" wires. Note that the neutral wire carries no current, as long as the loads are balanced. In other words, if both lamps draw the same current, there is no current flow in the neutral wire. However, if one of the lamps is turned off, then current must flow in the neutral wire. Observe that if a 240-volt appliance were connected across the two "hot" wires, no current would flow in the neutral wire. In practice, wiring systems are installed to maintain a reasonably balanced load condition in the 120-volt circuits. However, because lights are turned on or off in different combinations, a complete balance is impossible. Therefore, there is usually some current flowing in the neutral wire of a three-wire system.

HOW TO PLAN A WIRING INSTALLATION 3

BASIC PLANNING PROCEDURE

Basic planning consists of answering the question, "What is the purpose of this wiring system that is to be installed?" As an illustration, we might plan a wiring system to energize the lamps and appliances depicted in Figure 3-1, an example of an ordinary residential wiring installation. If the wiring system is to be installed while the house is under construction, it is called *new work*. On the other hand, we might plan a wiring installation to replace a single kitchen outlet with a multiple outlet capable of energizing several appliances without blowing ruses or tripping circuit breakers, as illustrated in Figure 3-2. Since the basic wiring system has been installed previously, the necessary modifications and additions are called *old work*. As another example, we might plan an old-work installation to correct inadequate wiring, without changing any of the existing outlets or fixtures. In other words, when an air conditioner is plugged into an outlet of an old-style wiring system, it may operate inefficiently or "stall" because of excessive current demand on the circuit. As we shall explain in greater detail subsequently, we can solve this problem by installing heavier circuit conductors.

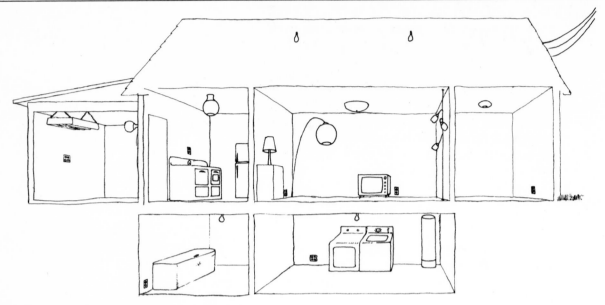

FIGURE 3-1   Placement of lamps and appliances in a typical
six-room house

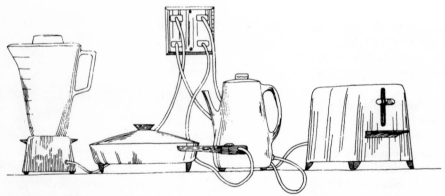

FIGURE 3-2   Four-outlet kitchen-appliance center

FUNDAMENTAL PARTS OF A WIRING SYSTEM

Any wiring system starts with a *service entrance,*
such as the one shown in Figure 3-3. Although a service entrance may take various forms, it always has
the purpose of conducting electrical power into a building through certain specified units of equipment. From
a practical viewpoint, a service entrance includes

service-drop wires running from the building to the public-utility line, *service-entrance* wires running from the outside of the building to the *service equipment* on the inside, a *kilowatt-hour meter*, a main *disconnect switch* to open the circuit between the service-entrance conductors and the wiring system in the building, main *circuit-breakers or fuses*, and a *ground connection* for the wiring system. In the example in Figure 3-3, the service-drop wires are supported by insulators and enter an *entrance head*. The kilowatt-hour meter is installed in a *meter socket*. A *service panel* houses the main disconnect switch, main circuit-breakers or fuses, and the ground connection.

A wiring system continues with *branch circuits*, as depicted in Figure 3-4. A branch circuit consists of the part of the wiring system that extends beyond the final automatic overload protective device. Each branch circuit is protected by its own circuit-breaker or fuse. The practical advantage of installing branch-circuit overload protective devices is that the entire wiring system does not "go dead" in case a branch circuit is overloaded. In turn, the service panel (distribution panel) contains *branch circuit-breakers or fuses* in addition to main circuit-breakers or fuses. We will find that some wiring systems also include *branch panels*, which contain additional branch circuit-breakers or fuses. Again, the service main disconnect switch may be housed separately from the branch circuit-breaker panel, as shown in Figure 3-5. In any disconnect arrangement, the switch is manually and externally operable. This means that we can pull or throw the switch by hand, without danger of touching any "live" wires or terminals.

Note that in Figure 3-5 branch circuits operate at 120 volts and at 240 volts. Heavy-current appliances are designed to operate at 240 volts, whereas lights and small appliances operate at 120 volts. We will find that branch circuits are also rated for maximum current demand; for example, a 120-volt branch circuit may be fused for either 15 amperes or 20 amperes. All branch circuits are connected in parallel to the service conductors. Therefore, the service installation must supply the sum of all the branch-circuit currents. As detailed subsequently, a service installation will be rated for 30, 60, 100, 150, or 200 amperes, maximum.

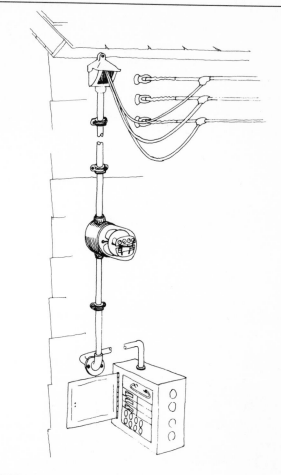

FIGURE 3-3   Service entrance

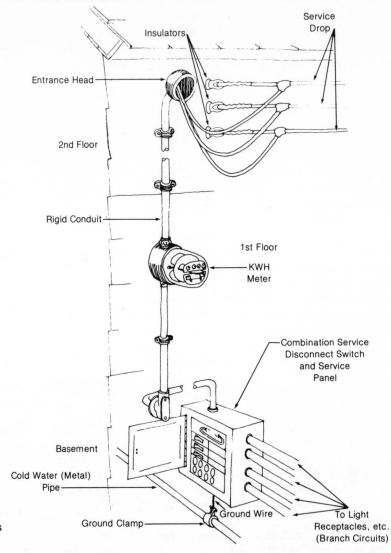

Service
Drop

Insulators

Entrance Head

2nd Floor

Rigid Conduit

1st Floor

KWH
Meter

Combination Service
Disconnect Switch
and Service
Panel

Basement

Cold Water (Metal)
Pipe

Ground Wire

Ground Clamp

To Light
Receptacles, etc.
(Branch Circuits)

FIGURE 3-4   The service panel feeds branch circuits

If the current rating of the service is exceeded for any reason, the main circuit breaker will be tripped, or the main fuses will be blown. In other words, the main overcurrent protective device might shut off the power to all of the branch circuits, without any one of the branch-circuit overcurrent protective devices being actuated.

Each branch circuit of a wiring system runs to various *boxes*. As an illustration, Figure 3-6 shows a

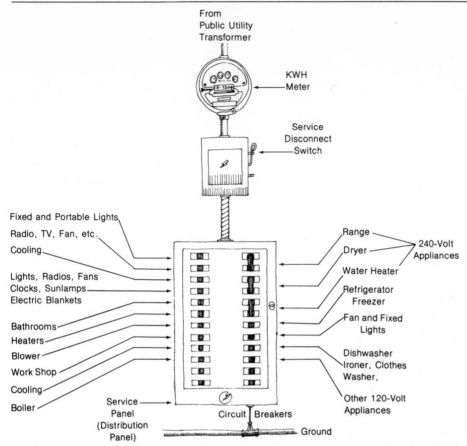

From
Public Utility
Transformer

KWH
Meter

Service
Disconnect
Switch

Fixed and Portable Lights
Radio, TV, Fan, etc.
Cooling

Lights, Radios, Fans
Clocks, Sunlamps
Electric Blankets

Bathrooms
Heaters
Blower
Work Shop
Cooling
Boiler

Service
Panel
(Distribution
Panel)

Circuit Breakers

Ground

Range
Dryer

240-Volt
Appliances

Water Heater

Refrigerator
Freezer

Fan and Fixed
Lights

Dishwasher
Ironer, Clothes
Washer,

Other 120-Volt
Appliances

FIGURE 3-5   Example of a service main-disconnect switch
installed in a separate housing

branch circuit that runs to a wall-bracket fixture, a
receptacle outlet, a light switch, a light fixture, an
automatic time switch, and a post-top lantern. Note
that the branch circuit also runs to a pair of junction
boxes; circuit connections must always be made inside
of boxes. This is a simple example of a branch circuit
that utilizes two conductors; it is typical of the installations encountered in old work. To anticipate subsequent discussion, let us add that new work generally
includes a third conductor, which is a continuous
ground wire through the system. Note also in Figure
3-6 that the wires are *color-coded* black and white in
this example. The black wire is "hot" and operates at

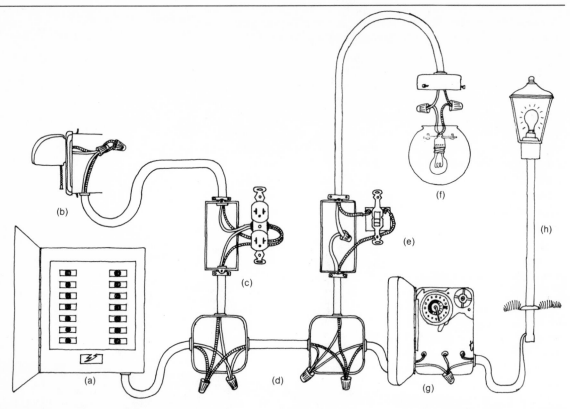

FIGURE 3-6   Arrangement of a simple wiring system
(a) Electric Service Entrance Panel: Main distribution panel which supplies current to branch circuits. Protects each circuit against danger of overload   (b) Wall Bracket Fixture: For kitchen or bathroom. Controlled by separate switch at base of fixture. Easily installed in an ordinary switch box   (c) Receptacle (d) Junction Box: Used to run wiring of a branch circuit in two or more directions. Occasionally used as a ceiling box for lights, or as an outlet box for receptacles   (e) Light Switch: Single-pole switch, used to control light from one point   (f) Light Fixture: Comes already wired   (g) Automatic Time Switch: Turn lights or power on and off at any set time. Note that the wiring is exactly the same as for any other switch   (h) Post-top Lantern: Wires lead from an ordinary switch box inside the house, through suitable underground conduit to opening at bottom of post, up the hollow post, and to the lamp terminals

120 volts above ground. The white wire is "cold" and is grounded at the service panel.

Observe that the simple wiring system depicted in Figure 3-6 consists of a single branch circuit from the service panel. In a complete wiring system, the branch circuit would be assigned a number. Thus, we might identify this part of the complete system as *branch circuit No. 1*. It follows from previous discussion that the total current demand of a branch circuit must not exceed the rating of the overcurrent protective device for that circuit. Otherwise, the overload will trip the branch circuit breaker or blow the branch fuses, and the entire branch will "go dead." It is both illegal and foolhardy to defeat an overcurrent protective device in any way, because of the resulting danger of fire to the building. In other words, the rating of an overcurrent protective device is determined by the size of wire that is installed in the branch circuit. Excessive current demand can cause a wire to become red-hot and set the insulation on fire.

If an appliance draws 10 amperes at 120 volts, it consumes 1200 watts; if an appliance draws 10 amperes at 240 volts, it consumes 2400 watts. However, the same size of wire can be used in the 120-volt branch circuit and in the 240-volt branch circuit, because the current rating (*ampacity*) of a conductor depends only on current demand, without regard to voltage. Next, if an appliance consumes 1000 watts for one hour, it consumes one kilowatt hour (kwh) of electrical energy. Thus, a kwh meter takes both current and voltage into account. Figure 3-7 illustrates the reading of a conventional kwh meter. The dials are read from left to right. Thus, the meter indicates 3392 kilowatt hours. Figure 3-8 lists the electrical energy consumption of various residential electrical equipment. We will find that a water heater may be energized from two circuits, one of which supplies lower-cost electricity through a time switch. The time switch automatically opens this circuit during peak-load hours.

Next, consider the current demands of the branch circuits depicted in Figure 3-9. This is a 100-ampere service panel. If each branch circuit were loaded to maximum capacity, and all branch circuits were in operation simultaneously, the total current demand would be 210 amperes, and the main breaker would be tripped. This demand does not even take the possibility of adding four more branch circuits for future

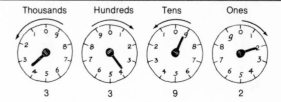

FIGURE 3-7   Reading a kilowatt-hour meter

ITEM	APPROX. KWH /MONTH	USE
Blanket (electric)	15	8 hr./day for 7 mo.
Clock	1½	25 hr./mo.
Dishwasher	25	1½ washings/day
Dryer (clothes)	50	10 hr./mo.
Fan (10-inch)	1	25 hr./mo.
Food Freezer	40	8 cu. ft.
Garbage Disposal	¾	4 min./day
Iron	6	12 hr./mo.
Ironer	10	10 hr./mo.
Lighting	65	
Mixer	¾	5 hr./mo.
Oil Furnace (excluding fan)	30	200-500 kwh/yr.
Radio	10	130 hr./mo.
Range	90	
Refrigerator	22	8 cu. ft.
Roaster	12	16 hr./mo.
Sandwich Grill	4	5 hr./mo.
Sewing Machine	1	
T.V.-Black-and-white	14	90 hr./mo.
T.V.-Color	27	90 hr./mo.
Toaster	3	3 hr./mo.
Vacuum Cleaner (upright)	2¼	6 hr./mo.
Vacuum Cleaner (tank)	3¼	6 hr./mo.
Washer (wringer)	2	12 hr./mo.
Washer (automatic)	3	12 hr./mo.
Water Heater	350	

FIGURE 3-8   Electrical energy consumption of various residential appliances

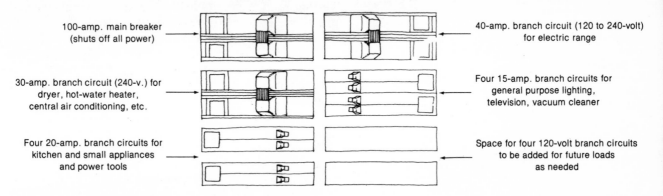

100-amp. main breaker (shuts off all power)

30-amp. branch circuit (240-v.) for dryer, hot-water heater, central air conditioning, etc.

Four 20-amp. branch circuits for kitchen and small appliances and power tools

40-amp. branch circuit (120 to 240-volt) for electric range

Four 15-amp. branch circuits for general purpose lighting, television, vacuum cleaner

Space for four 120-volt branch circuits to be added for future loads as needed

FIGURE 3-9   Example of a 100-ampere service-entrance panel

loads. Therefore, it might seem that this is an example of faulty planning. However, let us consider the current demands in somewhat greater detail. A 30-ampere branch circuit operating at 240 volts has a maximum power rating of 7200 watts. The 30-amp branch circuit in Figure 3-9 supplies a dryer, hot-water heater, and air-conditioner. These are fixed heavy appliances of the type that must be calculated at full rating in watts.

On the other hand, the 40-amp branch circuit in Figure 3-9 supplies an electric range. It is not likely that the oven(s) and all the burners will be in use at the same time. Therefore, the National Electrical Code (NEC) allows a *demand factor* to be applied in calculating service requirements. In this example, the range is not rated over 12 kw, and a reference value of 8 kw is stipulated. In turn, the current demand of this branch circuit is calculated as 33 amperes, approximately. Since it is extremely unlikely that every branch circuit will be loaded to its maximum capacity at the same time, the code stipulates a demand factor of 35 percent for the demand exceeding 3000 watts. This 35 percent demand factor applies to lighting, small appliances, and the laundry branch circuit. It excludes heavy fixed appliances. In turn, the four 20-amp branch circuits and the four 15-amp branch circuits are calculated at 5880 watts, instead of 16,800 watts. This results in a maximum current demand of 109 amperes, which is slightly greater than the 100-amp rating of the service. Finally, when we take into account the fact that the maximum possible load on certain branch circuits is less than their current rating, we arrive at a maximum current demand that is well within the rating of the 100-amp service.

The NEC permits another way of calculating the demand factor in planning service requirements. This method can be used only for a 100-amp service. It applies two demand factors to the total load that could be imposed on the branch circuits. The first 10 kw is calculated at 100 percent, and the remainder is calculated at 40 percent. If we apply this method to the branch circuits depicted in Figure 3-9, we start with a total load of 33.6 kw, of which 10 kw is computed at 100 percent, and 23.6 kw at 40 percent, giving 9.44 kw. Thus, we consider the total load to be 19.44 kw in this example. Since we consider this power value with respect to a 240-volt source, the maximum current demand is 81 amperes. Therefore, we arrive at a current demand that is well within the 100-amp rating of the service. We have ignored some minor details of load evaluation in this example. These are explained in detail in the prevailing National Electrical Code, which you should consult before finalizing plans for a wiring system. It is also essential to consult any applicable local codes.

With the foregoing principles in mind, the first requirement in planning a wiring system is to prepare accurate scaled wiring layouts, as discussed in the next chapter. After the layouts have been completed, take them to your local power company for checking, corrections, and recommendations. A service department is available for this purpose. Then, take the finalized layouts to your local planning commission, or equivalent governmental office, and determine whether you must take out a permit before starting the installation. To ensure that the completed installation will pass inspection, inquire at this time about any local regulations that may supersede the National Electrical Code. When purchasing materials for the installation, also make certain that they are approved by your local power company. In case of doubt, select materials that have been listed by the Underwriters Laboratories. Approved products carry a tag or stamp, such as those illustrated in Figure 3-10. Note that the power company generally supplies the meter and may furnish and install all wiring up to the meter.

FIGURE 3-10   Examples of Underwriters Laboratories' stamps and tags

TOOLS REQUIRED

Figure 3-11 shows basic tools used in electrical wiring work. If you are installing considerable wiring, a power bit such as the one shown in Figure 3-12 will save much labor. Electricians generally prefer to use a ship's auger bit, as illustrated in Figure 3-13. The advantage of this type of bit is reduced damage to the cutting edge of the bit in case a nail is accidentally struck. However, if several nails are struck in succession, even a ship's auger bit will be seriously damaged. If you are drilling masonry, you will require a carbide-tipped masonry drill like the one in Figure 3-14. When

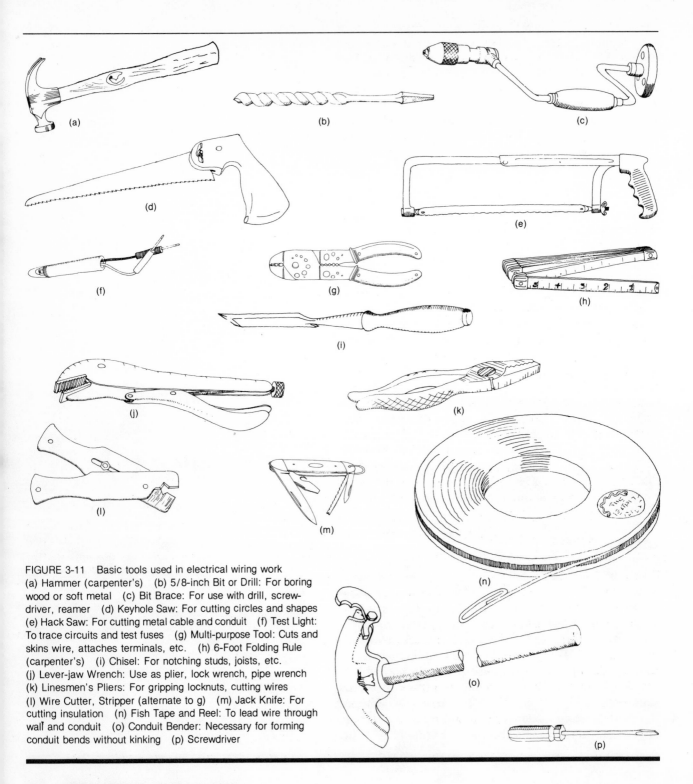

FIGURE 3-11   Basic tools used in electrical wiring work
(a) Hammer (carpenter's)   (b) 5/8-inch Bit or Drill: For boring
wood or soft metal   (c) Bit Brace: For use with drill, screwdriver, reamer   (d) Keyhole Saw: For cutting circles and shapes
(e) Hack Saw: For cutting metal cable and conduit   (f) Test Light:
To trace circuits and test fuses   (g) Multi-purpose Tool: Cuts and
skins wire, attaches terminals, etc.   (h) 6-Foot Folding Rule
(carpenter's)   (i) Chisel: For notching studs, joists, etc.
(j) Lever-jaw Wrench: Use as plier, lock wrench, pipe wrench
(k) Linesmen's Pliers: For gripping locknuts, cutting wires
(l) Wire Cutter, Stripper (alternate to g)   (m) Jack Knife: For
cutting insulation   (n) Fish Tape and Reel: To lead wire through
wall and conduit   (o) Conduit Bender: Necessary for forming
conduit bends without kinking   (p) Screwdriver

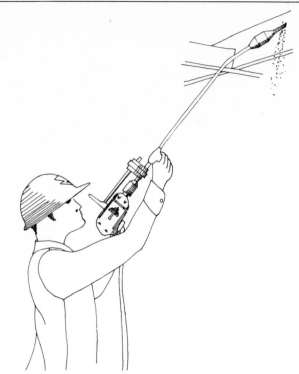

FIGURE 3-12 A power bit with extension for drilling ceiling joists

FIGURE 3-13 5/8'' single-spur, ship's auger power bit

hardwood boards must be lifted in old work, a special chisel is necessary to cut off the tongue on one of the boards. A suitable chisel can be fashioned by cutting a putty knife short and sharpening the edge. Cutting wallpaper in old work requires a razor blade. A small plumb bob on a chain is occasionally useful for sounding behind walls for obstructions in old work. When working in dark areas, you will need a flashlight or camp lantern.

BASIC FEATURES OF A FRAME BUILDING

An electrician needs to know the basic features of a frame building. Figure 3-15 shows a typical example of studs, joists, rafters, plates, fire block, head, sill, and bridging pieces. Joists are usually drilled as shown in Figure 1-12; studs are drilled in the same general manner. Plates must often be drilled, and fire blocks may

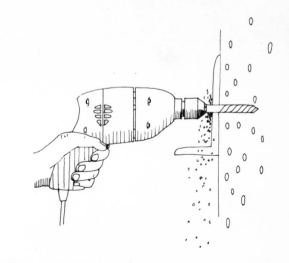

FIGURE 3-14 Carbide-tipped masonry drill

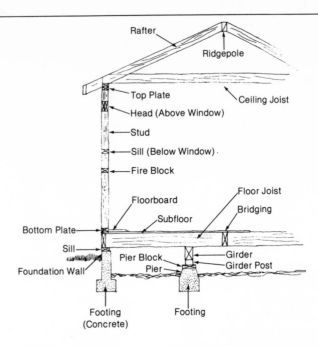

FIGURE 3-15   Basic features of frame construction

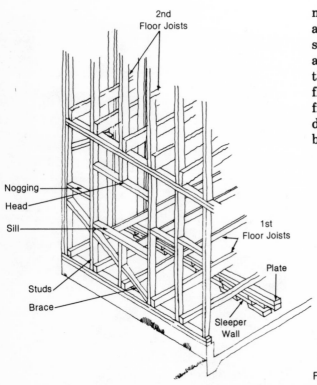

FIGURE 3-16   Additional basic features of frame construction

need to be drilled as well. Figure 3-16 depicts additional basic features of a frame building. Note that in two-story construction, the ceiling joists for the first floor are also the floor joists for the second floor. Similarly, the ceiling joists for the second floor may also be the floor joists for an attic. When wires run up from the first level to the second level, braces must often be drilled. Note that in old work, a brace, nogging, or fire block cannot be drilled. Therefore, these obstructions must be bypassed and the wires installed along some indirect and unobstructed route. Installation details are explained in subsequent chapters.

PRECAUTIONS IN PLANNING

Basic planning involves preparation of detailed wiring layouts, as noted previously. It is very poor practice to start a wiring project without a complete and approved layout. Preparation of a layout requires knowledge of the various kinds of switches, outlets, and fixtures that may be selected, plus an understanding of branch circuits. It is also necessary to know basic electrical wiring terms and the standard symbols used by electricians on wiring layouts. We have covered some of these facts in this chapter, and we shall explain additional practical data in the next chapter.

Among various precautions to observe are the following. (1) Always turn off the main switch and disconnect live appliances when rewiring, making repairs or alterations, or checking continuity. (2) Avoid making temporary wiring repairs that will not pass inspection. (3) Do not work on high-power transmission lines or the entrance wires to the kwh meter; this is the responsibility of the public utility. (4) When replacing fuses, never stand on a damp surface or a basement floor. (5) Never replace a blown fuse with a higher-rated fuse or attempt to defeat a fuse in any way. (6) Do not attempt to check circuits by short-circuiting between conductors or terminals with a screwdriver; use a voltage indicator such as depicted in Figure 3-11(f).

Finally, unless you have had previous experience in electrical wiring, do not attempt to plan a project until you have completed reading this book. In addition to National Electrical Code requirements, there are various tricks of the trade, established good practices, and certain procedures that are preferred on the basis of experience. These latter considerations are of particular importance in planning old-work projects. Statistically, more than half of the completed wiring projects included in a recent survey were found to be less than adequate a year later merely because of insufficient planning and checking before the wiring installations were started.

HOW TO DETAIL A WIRING LAYOUT 4

BASIC WIRING LAYOUTS

A basic wiring layout shows the outline or floor plan of a room or group of rooms with the locations of ceiling and wall fixtures, outlets, and switches indicated, as exemplified in Figure 4-1. Standard electrical wiring symbols indicate lights, outlets, and switches. A short list of symbols is given in Figure 4-2. Note that switches carry subscripts unless a single-pole switch is indicated. Three-way switches permit a light to be turned on or off from either of two locations. Four-way switches permit a light to be turned on or off from any of three or more locations. Convenience outlets may or may not be switched. Dotted lines running between lights and switches show the switch or switches that control particular lights. If a convenience outlet is switched, a dotted line is drawn between the outlet and its switch. Although a preliminary rough sketch may be prepared for a wiring layout, a scaled layout should be made on a planning chart like that in Figure 4-3 before approval is sought or material is ordered.

A completed wiring layout also indicates the number

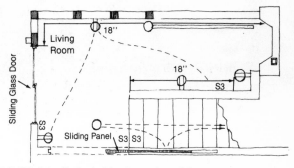

FIGURE 4-1   Typical wiring layout for a living room

○
⟩Ceiling Outlet
─○

─○ Wall Outlet

○L Ceiling Lighting Outlet

⊜ Duplex Convenience Outlet

⊜S Switch—Convenience Outlet

⊜WP Weatherproof Outlet

⊜R Electric Range

⊜D Electric Dryer

◌ 230-Volt Polarized Outlet

○━━━━━━━━ Fluorescent Light

◉ Special-Purpose Outlet

Ⓕ Ceiling Fan

─Ⓕ Wall Fan

Ⓙ Ceiling Junction Box

─Ⓙ Wall Junction Box

─Ⓢ─ Ceiling Pull Switch

Ⓒ Clock Outlet

Ⓣ Thermostat

Ⓖ Generator

Ⓜ Electric Motor

◍ Night Light

▣ Pushbutton

▭◗ Doorbell

─◁ Door Buzzer

─ⓇR Radio Outlet

─ⓉⓋ Television

S Single-Pole Switch

S₂ Double-Pole Switch

S₃ Three-Way Switch

S₄ Four-Way Switch

SWP Weatherproof Switch

FIGURE 4-2 Electrical wiring symbols

of the branch circuit that supplies a particular light or convenience outlet. As an illustration, Figure 4-4 shows a wiring layout for a residence with eight branch circuits called out. Note that each light and each outlet is accompanied by the number of the branch circuit with which it will be supplied. This portion of the

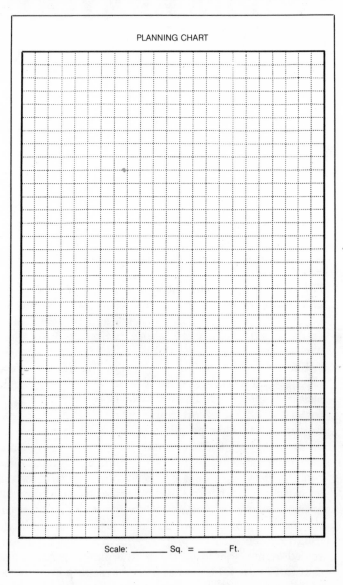

FIGURE 4-3 A scaled layout is made on a planning chart

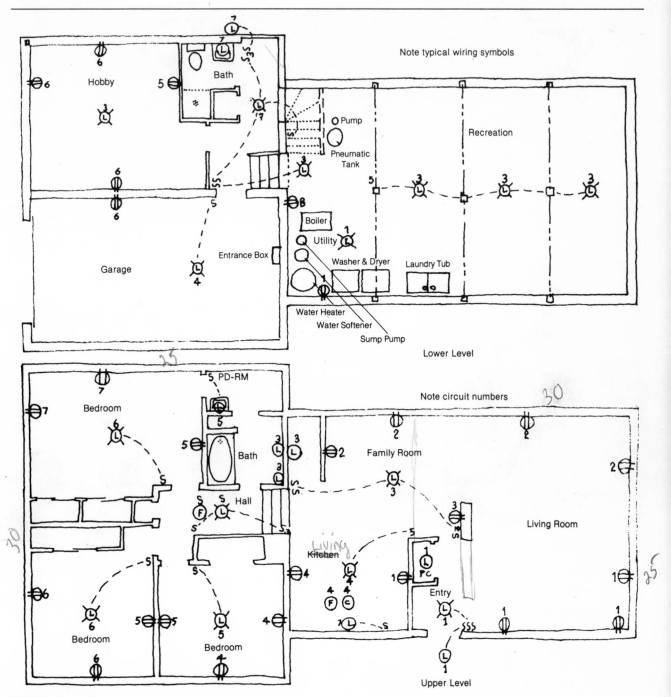

FIGURE 4-4 Typical wiring layout for a residence (a) Lower levels (b) Upper levels

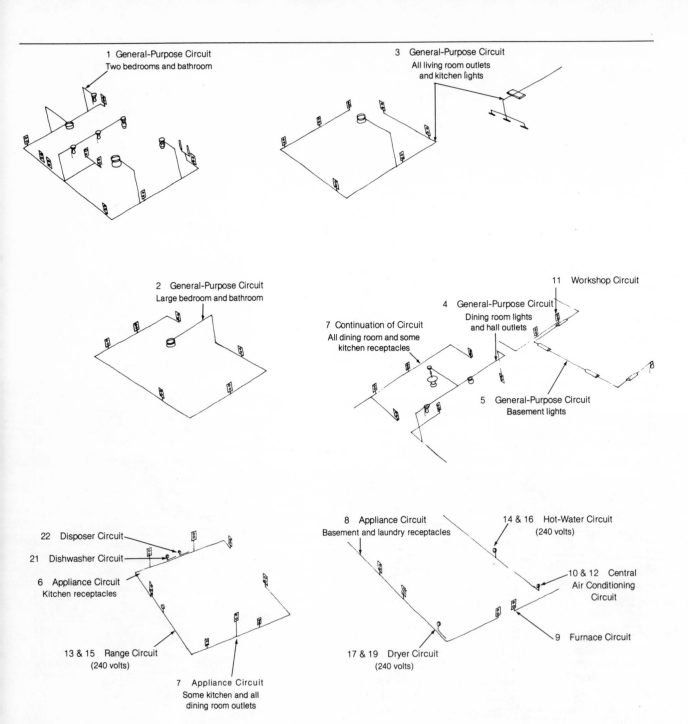

1 General-Purpose Circuit
Two bedrooms and bathroom

3 General-Purpose Circuit
All living room outlets
and kitchen lights

2 General-Purpose Circuit
Large bedroom and bathroom

11 Workshop Circuit

4 General-Purpose Circuit
Dining room lights
and hall outlets

7 Continuation of Circuit
All dining room and some
kitchen receptacles

5 General-Purpose Circuit
Basement lights

22 Disposer Circuit

21 Dishwasher Circuit

6 Appliance Circuit
Kitchen receptacles

8 Appliance Circuit
Basement and laundry receptacles

14 & 16 Hot-Water Circuit
(240 volts)

10 & 12 Central
Air Conditioning
Circuit

9 Furnace Circuit

13 & 15 Range Circuit
(240 volts)

17 & 19 Dryer Circuit
(240 volts)

7 Appliance Circuit
Some kitchen and all
dining room outlets

FIGURE 4-5 Floor sketches for electrical wiring in a six-room
house

planning procedure requires calculation of loads, demand factors, and practical considerations of wire runs. Figure 4-5 shows a breakdown on a branch-circuit plan, with the lights and outlets that are energized by each branch circuit. In this example, branch circuit No. 1 supplies seven fixtures and nine convenience outlets in two bedrooms and one bathroom. Branch circuit No. 2 supplies one fixture and six convenience outlets in a large bedroom and bathroom. Branch circuit No. 3 supplies two fixtures, three lights, and six convenience outlets in the living room and kitchen. Branch circuit No. 4 supplies three fixtures and two convenience outlets in the dining room and hall. Branch circuit No. 5 supplies five fixtures in the basement. Branch circuit No. 6 supplies four convenience outlets in the kitchen. Branch circuit No. 7 supplies six convenience outlets and three clock outlets in the dining room and kitchen. Branch circuit No. 8 supplies three convenience outlets in the basement and laundry. Branch circuit No. 9 supplies the furnace. Branch circuits No. 10 and 12 supply the central air-conditioning unit. Branch circuit No. 11 supplies two convenience outlets in the workshop (basement). Branch circuits 13 and 15 supply the electric range. Branch circuits 14 and 16 supply the hot-water heater. Branch circuits 17 and 19 supply the dryer. Branch circuits 18 and 20 (not shown in Figure 4-5) supply lights and appliances in the yard and outbuildings.

Another helpful breakdown of a branch-circuit plan is shown in Figure 4-6. Note that the branch circuits are grouped into general-purpose circuits, kitchen-appliance circuits, fixed-appliance circuits, and outdoor-building circuits. The general-purpose circuits may be rated at either 15 or 20 amperes. Kitchen-appliance circuits are rated at 20 amperes. Laundry appliances are supplied by a 20-ampere circuit. A fuel-fired furnace is supplied by a 15-ampere circuit. An air-conditioner circuit should be installed with an ampacity rating at least equal to the current rating of the range that is utilized. The same consideration applies to a hot-water heater circuit and a washer-dryer circuit. Outdoor-outbuilding circuits vary considerably in current demand from one installation to another. Therefore, it is necessary to add up the loads in these circuits to determine the required ampacity rating.

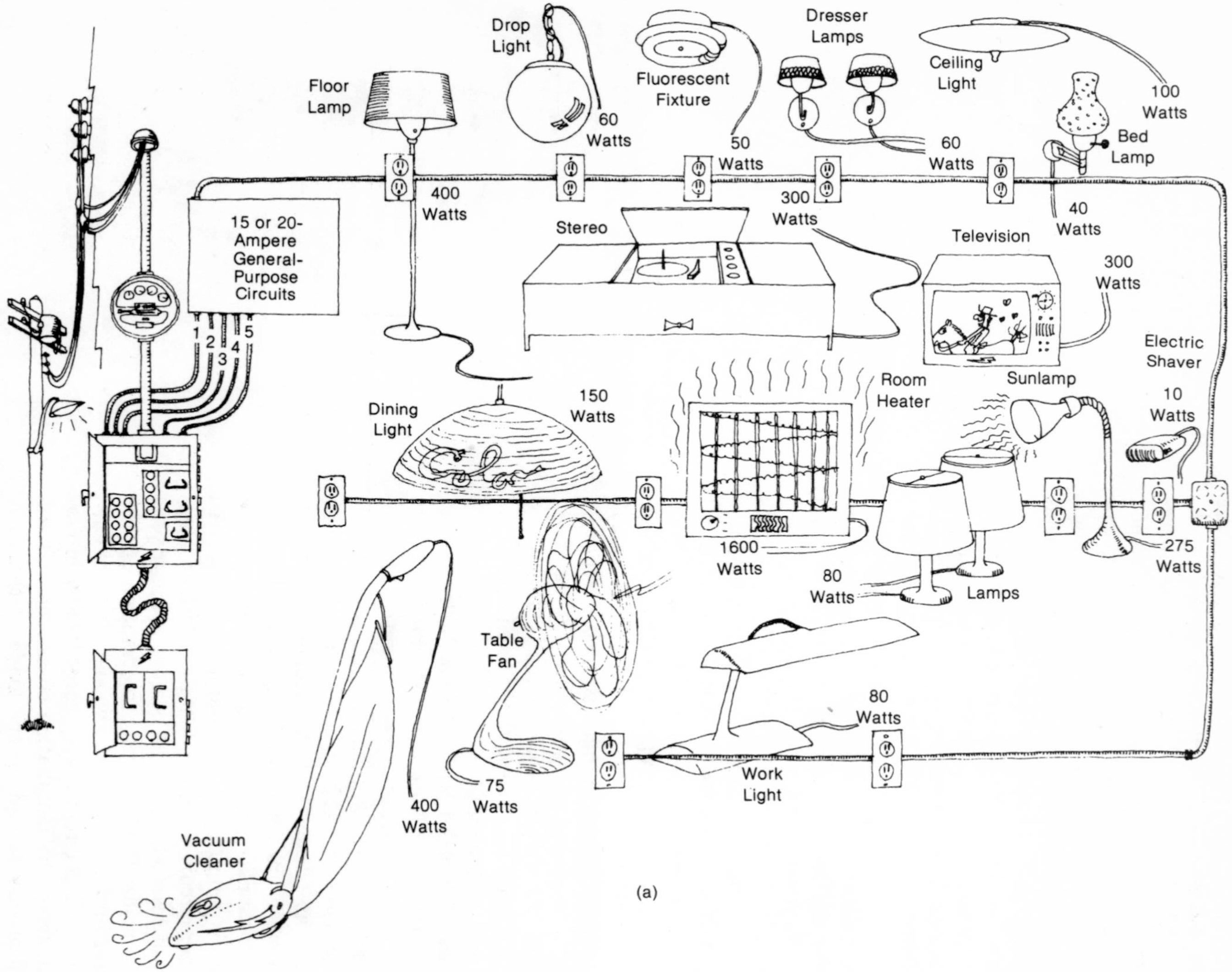

FIGURE 4-6 Practical example of branch circuiting (19 circuits) (a) General purpose circuits (1-5)

(a)

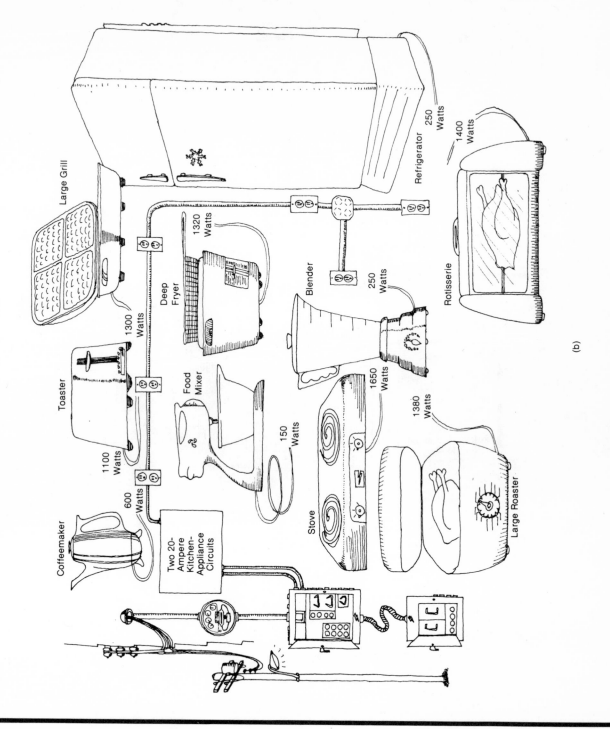

Large Grill

Refrigerator · 250 Watts

1400 Watts

Rotisserie

1320 Watts

Deep Fryer

Blender · 250 Watts

1300 Watts

Toaster

Food Mixer

1650 Watts

1100 Watts

150 Watts

1380 Watts

Stove

600 Watts

Coffeemaker

Two 20-Ampere Kitchen-Appliance Circuits

Large Roaster

(b)

FIGURE 4-5   (b) Kitchen appliance circuits (6-7)

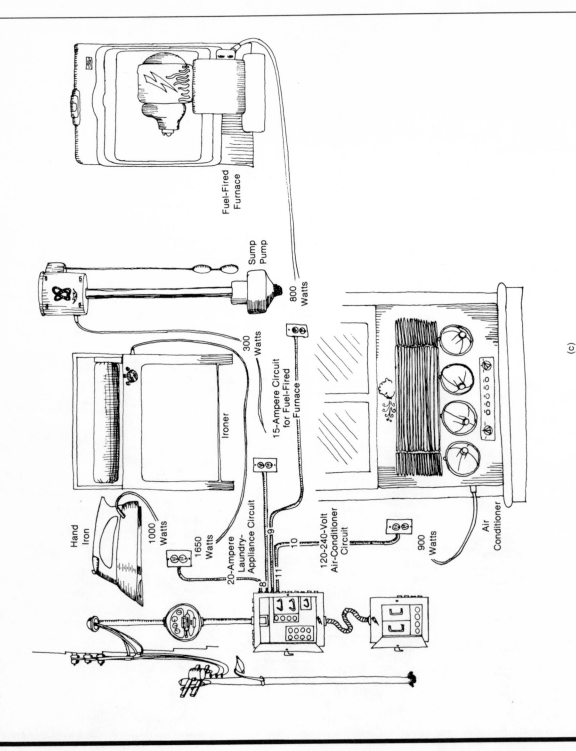

Fuel-Fired Furnace

Sump Pump

800 Watts

Ironer

300 Watts

15-Ampere Circuit for Fuel-Fired Furnace

Hand Iron

1000 Watts

20-Ampere Laundry-Appliance Circuit

1650 Watts

9

10

11

8

120-240-Volt Air-Conditioner Circuit

900 Watts

Air Conditioner

(c)

FIGURE 4-5 (c) Fixed appliance circuits (8–11)

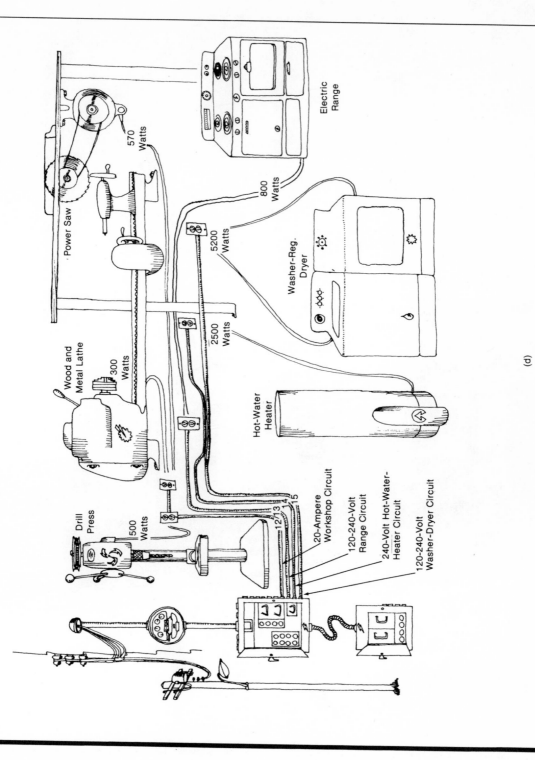

Power Saw

570 Watts

Electric Range

800 Watts

5200 Watts

Washer-Reg. Dryer

Wood and Metal Lathe

300 Watts

2500 Watts

Hot-Water Heater

Drill Press

500 Watts

12 13 14 15

20-Ampere Workshop Circuit

120-240-Volt Range Circuit

240-Volt Hot-Water-Heater Circuit

120-240-Volt Washer-Dryer Circuit

(d)

FIGURE 4-5   (d) Fixed appliance circuits (12-15)

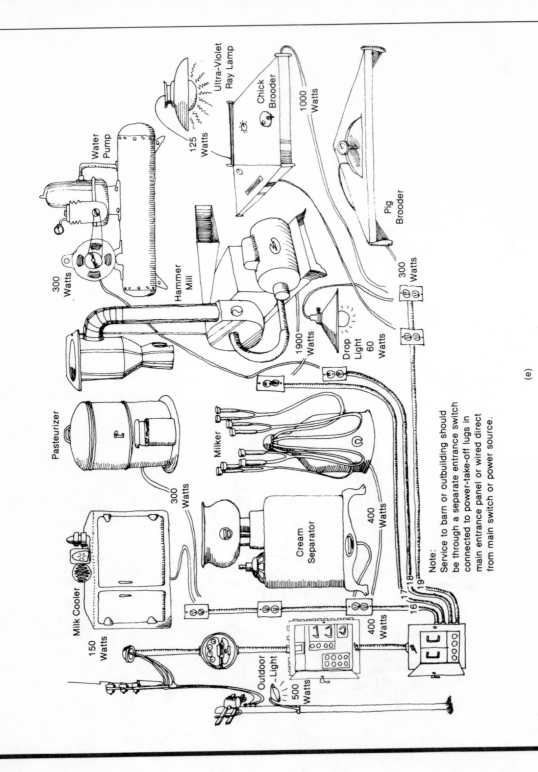

Water Pump

Ultra-Violet Ray Lamp

125 Watts

Chick Brooder

1000 Watts

Pig Brooder

300 Watts

300 Watts

Hammer Mill

1900 Watts

Drop Light 60 Watts

Pasteurizer

Milker

300 Watts

Cream Separator

400 Watts

Milk Cooler

150 Watts

Outdoor Light

400 Watts

500 Watts

Note:
Service to barn or outbuilding should
be through a separate entrance switch
connected to power-take-off lugs in
main entrance panel or wired direct
from main switch or power source.

16
17
18
19

(e)

FIGURE 4-5   (e) Outdoor-building circuits (16-19)

NATIONAL ELECTRICAL CODE REQUIREMENTS

When planning a wiring system, you should observe that the NEC requires installation of a convenience outlet every 12 feet of running wall space. Wall space afforded by fixed room dividers, such as free-standing bar-type counters must be included, so that no point along the floor line in any wall space is more than 6 feet, measured horizontally, from an outlet in that space, including any wall space 2 feet wide or greater, and the wall space occupied by sliding panels in exterior walls. Convenience (receptacle) outlets in any case must be spaced equal distances apart, insofar as practicable. Of course, if you wish to install more than the minimum number of outlets required by the code, this is your privilege. From the viewpoint of circuiting, two 20-ampere appliance circuits must be installed for the kitchen, dining room, and one 20-ampere appliance circuit for the laundry, independent of lighting fixtures. In other words, lighting fixtures must not be supplied by the foregoing 20-ampere appliance circuits. However, the same 20-ampere appliance circuits may be run to the kitchen, dining room, and laundry, if desired.

A kitchen wiring layout must include one convenience outlet in any counter space wider than 12 inches. If a kitchen space is separated by a range top, refrigerator, or sink, it is considered an individual space that must be provided with a convenience outlet. In any case, it is good practice to plan for an outlet every 4 feet of counter space, although the NEC might permit a greater spacing. As noted above, two 20-ampere circuits must be installed in a kitchen, and it is good practice to supply one outlet from one circuit, and to supply the next outlet from another circuit, and so on. This wiring layout minimizes the chances of tripping a circuit breaker or blowing a fuse as a result of overloading one of the kitchen appliance circuits. Similarly, good planning places the outlets on any floor on more than one circuit. Not only does this arrangement lessen the chances of overload, but in case of accidental overload (which does occur on occasion), the entire floor will not be blacked out. As a general rule of good planning, put most of the branch circuits where the load is the heaviest.

In areas where several small appliances or lamps may need to be supplied from adjacent outlets, it is good planning to install a multiple-outlet (plug-in strip) such as the one shown in Figure 4-7(a). This foresight will avoid the hazard of a future "octopus" tangle of multiple-outlet taps and extension cords. See Figure 4-7(b). Note that the NEC imposes only minimum requirements for a wiring installation; these requirements should often be exceeded in planning for utility, convenience, efficiency, and future needs. The NEC requires provision of at least three watts of electrical power for each square foot of floor area when planning lighting circuits. Additional power must be provided for appliance circuits. Note that floor area is calculated from the outside dimensions of a building. Open porches, garages, and unused or unfinished spaces do not count as floor area. However, if such spaces are adaptable for future use as dwelling space, they must be considered floor area when planning a wiring system. Although not required, the NEC recommends provision of one branch lighting circuit for each 500 square feet of floor area.

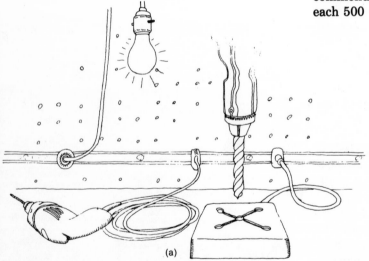
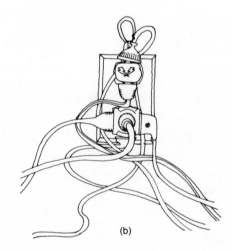

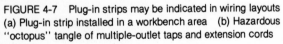

FIGURE 4-7   Plug-in strips may be indicated in wiring layouts (a) Plug-in strip installed in a workbench area   (b) Hazardous "octopus" tangle of multiple-outlet taps and extension cords

Again, the NEC does not require a separate circuit for each type of heavy fixed appliance. However, it is general installation practice to provide a separate circuit for an electric range or separate oven, a dishwasher, an automatic clothes washer, a dryer, a garbage disposal, or a water heater. It is also general practice to install a separate circuit for any fixed appliance, such as a bathroom heater, if it is rated for 1000 watts or more. Any motor rated at 1/8 horsepower or more, such as a furnace blower, is customarily provided with a separate circuit. Figure 4-8 depicts a typical branch-circuit installation for four heavy fixed appliances. Details of installation procedure are explained subsequently. As noted previously, most electric ranges operate on 120/240-volt circuits. High-heat burners operate on 240-volts, whereas low-heat burners operate on 120 volts. From the veiwpoint of planning and layout, both voltages are supplied by the same range circuit (using three wires).

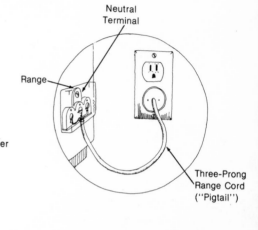

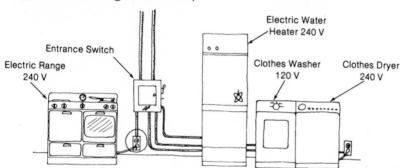

FIGURE 4-8   Typical branch circuiting for three 240-volt and one 120-volt heavy fixed appliances

SPLIT-OUTLET CONSIDERATIONS

To minimize the possibility of blowing fuses or tripping circuit breakers, good planning requires supplying multiple outlets by different circuits. As an illustration, Figure 4-9 shows a duplex receptacle; it permits two appliances to be plugged in at the same outlet box. Metal links connect the two outlets in parallel. In turn, the receptacle is easily wired into a single branch circuit. Note that if two appliances with comparatively large current demands were plugged into the same receptacle, the branch circuit would be overloaded in some cases, with the result that the overcurrent protective device would be actuated. This nuisance can be prevented by connecting each of the outlets to a different branch circuit. To anticipate subsequent discussion, this is called split-outlet installation; it involves breaking out the parallel linkages with a screwdriver,

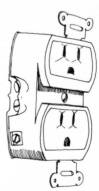

FIGURE 4-9   A duplex receptacle is supplied with parallel-connecting links

and wiring the receptacle into two branch circuits. Therefore, when planning a wiring layout with split outlets, two branch circuits will be shown running to the same duplex receptacle. It is even more important to employ split outlets when a four-outlet box is installed, such as that in Figure 4-10.

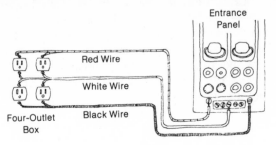

FIGURE 4-10   Basic split-circuit arrangement for a four-outlet box

When detailing a wiring layout, you may wish to indicate switch control of certain split outlets. It is customary to place the switch in the circuit that supplies the lower outlet in a duplex receptacle, leaving the upper outlet "hot" at all times. Switch control of outlets should be planned for floor lamps, but "hot" outlets should be planned for electric clocks. Old-work planning, in particular, occasionally poses a problem of an insufficient number of branch circuits. In such a case, the most economical solution is to lay out an add-on fuse panel or circuit-breaker panel connected to the power take-off lugs and the neutral strip in the service-entrance panel, as depicted in Figure 4-11. The arrangement shown provides two additional branch circuits. Note in passing that if the final demand factor indicates that the existing service conductors will be overloaded, they must be replaced with larger conductors.

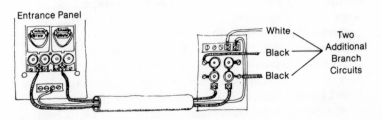

FIGURE 4-11   Basic arrangement for an add-on fuse panel

TYPICAL SINGLE-FAMILY RESIDENCE

Planning a wiring layout for a typical single-family residence involves the following considerations. The floor area in thes example is 1500 square feet, not counting an unoccupied cellar, an unfinished attic, and open porches. The general lighting load is calculated at 3 watts per square foot, or 4500 watts. A 4500-watt power demand corresponds to 39.1 amperes at 115 volts. Therefore, three 15-ampere branch circuits, or two 20-ampere branch circuits. Finally, the laundry current demand of small appliances will be supplied by two 20-ampee branch circuits. Finally, the laundry load requires a 20-ampere branch circuit, and the 12kw range will require a separate circuit. In summary, at least six and possibly seven branch circuits will be required in this wiring layout.

To calculate the ampacity of service that is required, we proceed as follows. The general lighting load is 4500 watts, the small appliance load is 3000 watts, and the laundry load is 1500 watts. Thus, the power demand, not counting the range, is 9000 watts. Demand factors may be applied to this power value as follows. The first 3000 watts is calculated at 100 percent. The remaining 6000 watts may be calculated at 35 percent, or 2100 watts. Thus, the total power demand in accordance with the demand factors is 5100 watts, not counting the range. Although the range is rated for a maximum power of 12 kw, the demand factor permits this load to be calculated at 8000 watts. Therefore, the total power value must be supplied by the service is 13,100 watts. We divide this power value by 230 volts, and arrive at a service current demand of 57 amperes. Although this figure is well within a 100-ampere limit, the net computed load exceeds 10 kw. In turn, the NEC requires installation of a 100-ampere service in this example.

SINGLE-FAMILY RESIDENCE WITH AIR CONDITIONING

Next, let us consider the addition of air conditioning to the foregoing example. A 6-ampere, 230-volt room air conditioner and three 12-ampere, 115-volt air conditioners will be stipulated. We arrived previously at a service requirement of 57 amperes, to which we must add the air-conditioning load. When a 230-volt service is utilized (which is always the case for heavy fixed appliances), we calculate the current demand in each feeder and base our conclusion on the heavier of the two demands. To illustrate this point, refer to Figure 4-11; note that the service feeders consist of two black wires and a white wire. The white wire is called the neutral conductor. There is a 230-volt potential between the two black wires, and there is a 115-volt potential between either one of the black wires and the white wire. It is evident that the 115-volt branch-circuit loads might be more or less unequal. Therefore, one of the black wires in the service feeder could have a higher current demand than the other black wire.

To continue with our current-demand calculations, the service requirement of 57 amperes that we previously calculated is assumed to be equally divided in each of the black wires. To this value, we add the 6-ampere current demand of the 230-volt air conditioner. Since a 230-volt appliance draws equal currents from each of the black wires, we arrive at a current demand of 63 amperes in each of the black wires. Next, we add the 12-ampere current demand of the two 115-volt air conditioners. One of these will be installed in the first 115-volt branch circuit, and the other will be installed in the second 115-volt branch circuit. Therefore, we will add 12 amperes to the current demand in each black wire, or a total of 75 amperes. Now, we have one more 115-volt air conditioner to account for, and it must be supplied by a single branch circuit. Therefore, the black wires will have an additional 12-ampere current demand, bringing us to a total of 87 amperes in this heavier-loaded black wire. The NEC requires that an additional current demand be included which is calculated as 25 percent of the current demand of the largest motor in the wiring system. Therefore, we must add 3 amperes to our previous total, giving a grand total of 90 amperes per line. We conclude that, although a 100-ampere service could be installed, no future appliances could be accommodated. Therefore, good planning practice dictates selection of a 150-ampere or 200-ampere service.

ELECTRIC HEATING

When planning a wiring layout, consideration should be given to the options of electric heating. Electric heating has the advantages of clean, silent, and flameless operation. Uniform temperatures are maintained automatically by thermostat control, and space is not required for fuel storage or for a heating plant. Lower power rates are often provided by public utilities for residential heating. An entire building can be heated electrically, or one room, or an enclosed porch, and so on. When planning several heating units or a heating system, 230-volt circuits should be employed. Small single heaters can be operated from a 115-volt circuit. Do-it-yourselfers generally restrict installations to 115-volt supplementary heating units. Typical equipment is illustrated in Figure 4-12. During the planning stage, it is essential to consult with the local public utility to determine if the building will be approved for electric heating.

OUTDOOR WIRING LAYOUTS

Most outdoor wiring installations are made with underground cable. For example, Figure 4-13 shows a pictorial plan for a post-lamp and spike-light circuit. Inclusion of a time switch permits outdoor lights to be

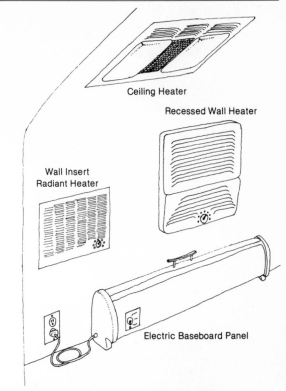

FIGURE 4-12   Typical supplementary electric heating units

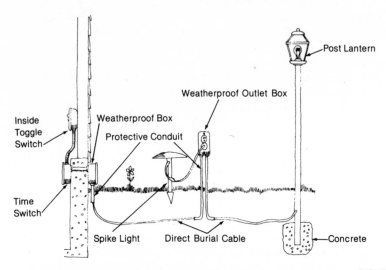

FIGURE 4-13   Post-lamp and spike-light circuit

automatically switched on and off at chosen hours. Alternatively, a photoelectric switch can be used, in order to turn the circuit on or off depending upon the prevailing light level. Underground cable is also installed for outdoor outlets located on a garage or other outbuilding, as shown in Figure 4-14.

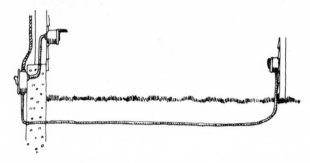

FIGURE 4-14   Example of outdoor outlets with an underground cable run

It is good planning practice to specify an outdoor outlet in the vicinity of any area where power tools or appliances such as lawn mowers, hedge clippers, or barbecue equipment may be operated. If preferred, an outdoor wiring installation can be planned for overhead wiring, as illustrated in Figure 4-15.

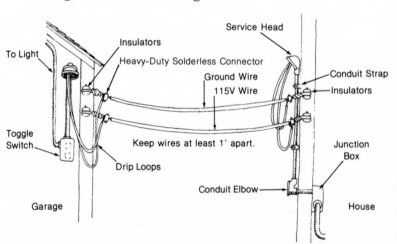

FIGURE 4-15   Outdoor wiring circuit with overhead wiring

Outdoor wiring circuits may also be provided for energizing heating tape. Figure 4-16 shows examples of heating tape use. Heating tape keeps outdoor stairs and walks free from sleet, prevents exposed pipes from freezing, melts ice in gutters, and so on.

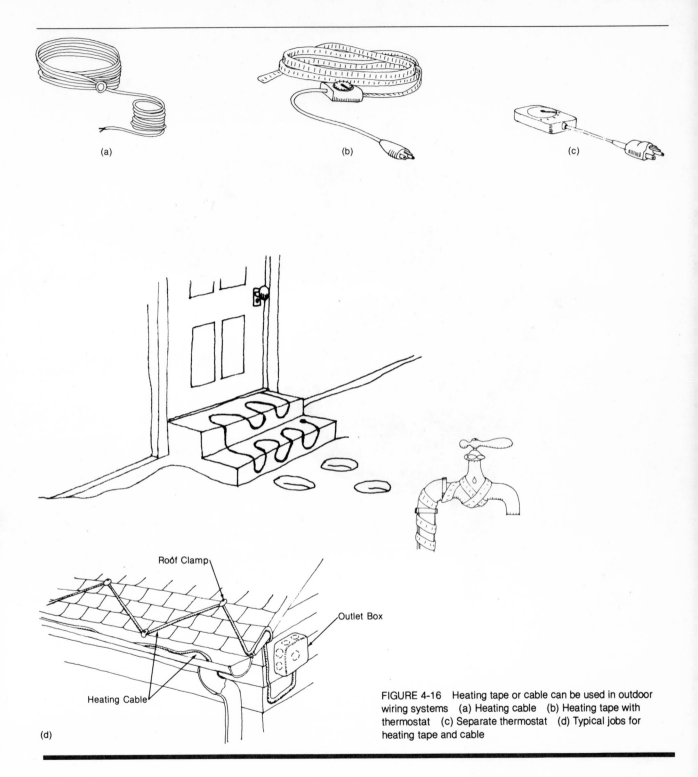

(a)

(b)

(c)

Roof Clamp

Outlet Box

Heating Cable

(d)

FIGURE 4-16 Heating tape or cable can be used in outdoor wiring systems (a) Heating cable (b) Heating tape with thermostat (c) Separate thermostat (d) Typical jobs for heating tape and cable

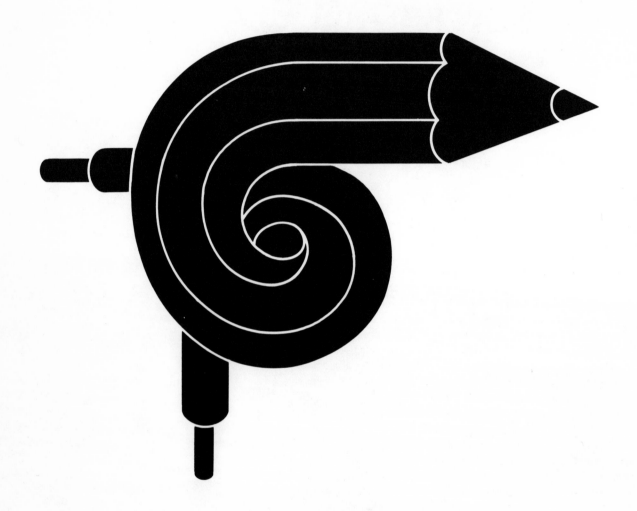

HOW TO MAKE A WIRING LIST OF MATERIALS 5

COPPER AND ALUMINUM CONDUCTORS

Two of the most widely used types of electrical conductors are shown in Figure 5-1. These are the indoor-type plastic sheathed cable and the dual-purpose plastic sheathed cable. They are also called nonmetallic sheathed cable or Romex. The NEC refers to the indoor type as NM cable, and to the dual-purpose type as NMC cable. Indoor-type plastic sheathed cable is available with either copper or aluminum conductors. It has a tough, flexible outer jacket which is flattened in shape and ivory colored. This cable can be used for all types of wire runs indoors. The dual-purpose plastic sheathed cable can be installed indoors or outdoors, and can be buried underground. It resists moisture, acids, and corrosion. This cable can be run through masonry or between studding. It is available with copper conductors only.

Figure 5-2 depicts flexible armored cable. It is installed in dry locations of exposed runs on wall or ceiling surfaces, or for concealed runs in hollow spaces of wall, floors, or ceilings. This type of armored cable can be embedded in wall or ceiling plaster, if the location is

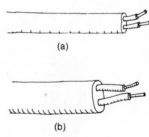

FIGURE 5-1   Two widely used types of wiring cable   (a) Indoor-type plastic sheathed cable   (b) Dual-purpose plastic sheathed cable

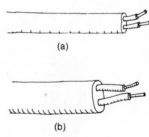

FIGURE 5-2   Flexible armored cable

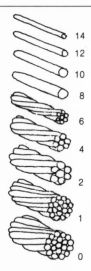

FIGURE 5-3 Actual sizes of several gauge numbers of copper conductors

not excessively damp. It cannot be installed in damp indoor locations, outdoors, or underground. The wires are encased in a heavy steel cover, which is flexible. This type of cable (or nonmetallic sheathed cable) is often used for extensions of existing conduit systems in old work. Armored cable must be used with steel switch boxes or junction boxes, and cannot be used with plastic or porcelain boxes, as explained in greater detail subsequently. This type of cable is commonly called BX, and is termed AC cable by the NEC.

Electrical conductors are specified in diameter in AWG gauge numbers. Figure 5-3 shows the actual sizes and corresponding gauge numbers of conductors used in typical wiring systems. Gauge numbers refer to the bare wire, without insulation. Electricians generally use a wire gauge such as the one shown in Figure 5-4 to check on wire sizes. Note that aluminum has higher resistance than copper, although aluminum is lighter and more easily bent. Because of its higher resistance, a larger diameter aluminum wire must be used in place of copper wire. When the NEC specifies wire sizes, the reference is to copper wire. If aluminum wire is used instead of copper, its diameter should be two sizes larger. For example, if the NEC required No. 12 copper wire in a particular installation, No. 10 aluminum wire will be required in the same installation. Or, if No. 14 copper wire is specified, No. 12 aluminum wire must be used in its place.

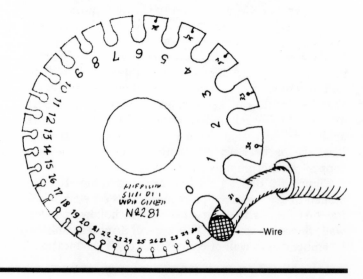

FIGURE 5-4 An American standard wire gauge

SERVICE-ENTRANCE CONDUCTORS

A 30-ampere service (such as might be installed in a roadside fruit stand) uses No. 8 conductors in a 2-wire 120-volt service. A 30-ampere service panel is utilized, and the service provides only limited ampacity for lighting and a few of the smaller appliances. A 60-ampere service, however, employs No. 6 conductors in a 3-wire 120/240-volt service. It provides ampacity for lighting and portable appliances including a range, and a dryer or a hot-water heater, but no other major appliances. Again, a 100-ampere service uses No. 2 or No. 3 conductors in a 3-wire 120/240-volt service with a 100-ampere service panel. The service conductors utilize type RHW insulation, which is suitable for installation in either wet or dry locations. Service-entrance cable is called SE cable, and if it is suitable for underground installation it is called USE cable. A 100-ampere service can be installed in residences up to 3000 square feet in floor area. A typical service-entrance cable is depicted in Figure 5-5. Three-wire service is generally installed with three separate RHW-insulated conductors.

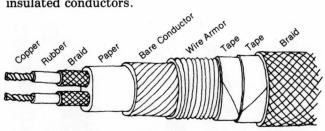

FIGURE 5-5   Typical service-entrance cables

A 150- or 200-ampere service utilizes No. 1/0 or No. 3/0 conductors with RHW insulation. Figure 5-3 shows that a single No. 0 conductor is denoted as 1/0; if three No. 0 conductors are combined (as by being twisted together), the cable is denoted as 3/0. A 150-ampere service is adequate for lighting and portable appliances including ironer, roaster, rotisserie, refrigerator, 12-kw range, 8700-watt clothes dryer, and 5 kw (3 to 5 tons) of air conditioning plus an appliance load up to 5500 watts. However, if electrical house heating is to be installed, a 200-ampere service will be required. Note that a two-wire service employs a black wire and a white wire; a three-wire service utilizes one black, one red, and one white wire.

INTERIOR WIRING REQUIREMENTS

Most interior wiring is installed with plastic sheathed cable. The conductor size required by the NEC depends upon the current demand of the circuit. For example, No. 14 copper wire has a maximum ampacity of 15 amps; No. 12, 20 amps; No. 10, 30 amps; No. 8, 40 amps; No. 6, 55 amps. Figure 5-6 lists the wire sizes required for use with various appliances, with the power consumption, current demand, number of conductors, and recommended fuse rating for each type of appliance. When making up a material list, remember to multiply the length of the run by three for a three-conductor circuit. It is a good idea to order a little more wire than is indicated by the length of the runs, to allow for connections in boxes with necessary slack, unanticipated bends, and so on.

APPLIANCE	WATTS	VOLTAGE	WIRES Qty.	WIRES Size	FUSE (AMP.)
Electric Range	12000	115/230	3	# 6	50-60
Dishwasher	1200	115	2	#12	20
Garbage Disposer	300	115	2	#12	20
Refrigerator	300	115	2	#12	20
Home Freezer	350	115	2	#12	20
Automatic Washer	700	115	2	#12	20
Automatic Dryer	5000	115/230	3	#10	30
Rotary Ironer	1650	115	2	#12	20
Water Heater	—Check with utility company—				
Power Workshop	1500	115	2	#12	20
Television	300	115	2	#12	20
20,000 Btu Air Conditioner	1200	115	2	#12	20
Heating Plant	600	115	2	#12	15-20
Central Air-Conditioning System	—Check with utility company—				
Space Heating	—Check with utility company—				

FIGURE 5-6   Wire sizes required by various appliances

Various cable components must also be included in your wiring list of material. When nonmetallic cable is installed, cable straps such as those shown in Figure

5-7 are used to support the cable along joists and studdings. The cable should be strapped at 4½ foot intervals, and within 6 to 12 inches of every outlet box. Straps are not required in old work, where cable is fished through floors or walls; however, straps are mandatory in new work. Outlet box connectors are also required, as depicted in Figure 5-7(d), when cable is run through a knockout in the box. It is also helpful to have a cable ripper to remove cable insulation, as shown in Figure 5-7(c), although an electrician's knife can be used instead of a cable ripper, if desired. Note that in some areas cable clamps may be replaced by large staples, and the installation will pass inspection.

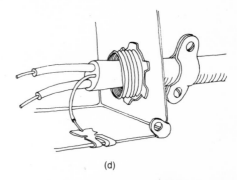

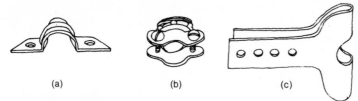

(a)   (b)   (c)   (d)

FIGURE 5-7 Components used with nonmetallic cable
(a) Cable strap (b) Cable-to-box connector (c) Cable ripper
(d) Nonmetallic cable (with bare ground wire) secured through knockout with connector

When armored cable, as in Figure 5-8(a) is used, a protective fiber bushing (Figure 5-8(b)) is required at each end of a cable section. This bushing prevents the raw metal edge from cutting into the conductor insulation. Outlet box connectors are required (Figure 5-8(c)) when the cable is run through a knockout in the box. Armored cable must be supported with straps or staples, as in Figure 5-8 (d) and (e). The cable must be

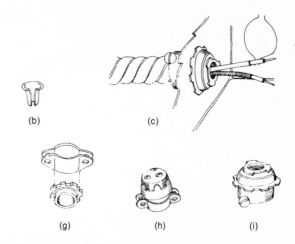

(a)   (b)   (c)

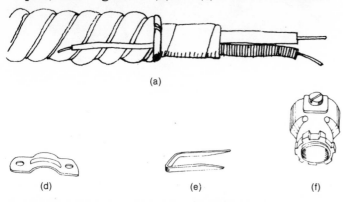

(d)   (e)   (f)   (g)   (h)   (i)

FIGURE 5-8 Components used with armored cable
(a) Armored (BX) cable, with bonding wire (b) Fiber bushing
(c) BX cable secured through knockout with connector

(d) Cable strap (e) Cable staple (f) Duplex connector
(g) 90° angle connector (h) End fitting (i) Cable-to-box connector

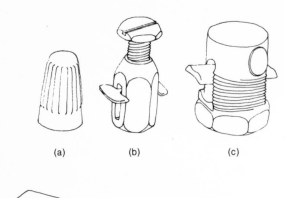

(a)   (b)   (c)

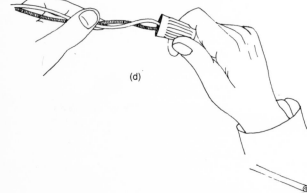

(d)

supported at 4½-foot intervals, and within 6 to 12 inches of every box. However, straps or staples are not required in old work where the cable is fished through walls or ceilings. A duplex connector (Figure 5-8(f)) is required when securing two sections of cable through one knockout. Alternatively, a 90° angle connector (Figure 5-8(g)) may provide easier installation. If you need to change from a 2-wire or a 3-wire armored cable to open wiring, you will need an end fitting (Figure 5-8(h)).

Your wiring list of material must include a number of solderless connectors, such as those depicted in Figure 5-9. Connectors with plastic caps, as shown in Figure 5-9(a), are used chiefly in boxes, where there is no strain on the wires. When tapping a line where there is strain on the wires, a heavy-duty connector is utilized, such as the one in Figure 5-9(b). Service-entrance connections are made with heavy-duty solderless connectors; see Figure 5-9(c). Installation procedures will be explained subsequently. Although it is impractical to determine precisely how many connectors will be required in a wiring installation, a review of your wiring layout will lead you to a reasonable approximation in compiling a list of material.

Type A

(e)

Type B

(f)

FIGURE 5-9   Types of solderless connectors   (a) Plastic-cap connector   (b) Tap connector   (c) Service-entrance tap connector   (d) Connection made with plastic-cap connector   (e) Type A tap connector in use   (f) Type B (service-entrance) tap connector in use

THIN-WALL AND RIGID CONDUIT

A wiring layout may specify installation of thin-wall conduit (EMT or electrical metallic tubing), such as that shown in Figure 5-10. Some local codes require the use of EMT. It provides protection for the wires and acts as a ground for the entire wiring system.

Steel Tubing

FIGURE 5-10   Thin-wall conduit

EMT is comparable to water pipe, except that it has thinner walls. Installation is made with special threadless fittings illustrated in Figure 5-11. EMT is supplied in 10-foot lengths and may be used for concealed or exposed work in wet or dry locations. However, EMT cannot be buried in cinders or in cinder concrete. Wires are drawn through the conduit after it is installed. The wires are single-strand, insulated with thermoplastic insulation, and unjacketed. Wires are color-coded black, white, and red. The required diameter of conduit depends on the size of the wires and the number of wires, in accordance with Table 5-1.

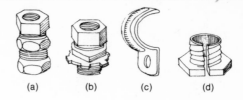

FIGURE 5-11   Threadless fittings used with EMT   (a) Coupling to join sections of EMT   (b) Conduit-to-box connector   (c) Conduit-to-box connector   (c) Conduit strap   (d) Adapter to use EMT in fittings designed for rigid conduit

CONDUIT SIZE AND AMPACITY OF WIRES IN CONDUIT
Number of Wires (1 to 9) to be Installed in Conduit.
(Varies per Local Code)

Wire Size	Ampacity	½-inch Conduit	¾-inch Conduit	1-inch Conduit	1¼-inch Conduit
14	15	4	6	9	9
12	20	3	5	8	9
10	25	1	4	7	9
8	35	1	3	4	7
6	45	1	1	3	4
4	60	1	1	1	3
2	95	1	1	1	3

WIRE SIZES AND AMPACITIES FOR 115-VOLT CIRCUITS

Wire Size	Max. Fuse Amps.	5A 575W	10A 115W	15A 115W	20A 2300W	25A 2875W	35A 4025W
14	15	90	45	30			
12	20	140	70	47	35		
10	25	220	110	75	60	45	
8	35	360	175	125	90	75	55
6	45	560	280	190	150	120	85

Distance (feet) One Way for Amps and Watts

WIRE SIZES FOR MOTOR CIRCUITS
Distance from Motor to Fuse Box or Meter

H.P.	0 to 50 Ft. 115V	230V	50 to 100 Ft. 115V	230V	100 to 150 Ft. 115V	230V	150 to 200 Ft. 115V	230V
¼	14	—	12	—	12	—	10	—
⅓	14	14	12	14	12	14	10	14
½	14	14	10	14	10	14	8	14
¾	12	14	10	14	8	14	6	12
1	12	14	8	14	6	12	6	12
1½	10	12	8	12	6	8	4	6
3	8	14	4	10	4	8	2	8
5	6	12	2	8	0	6	0	6

TABLE 5-1   Conduit and Wire Data

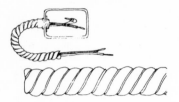

FIGURE 5-12   Flexible conduit, or Greenfield

Note that flexible conduit (Greenfield) may be installed instead of EMT in old work. Greenfield (Figure 5-12) is similar to armored cable, except that it contains no conductors. Wires are drawn through Greenfield after it is installed. The same fittings are utilized as for armored cable. Greenfield is recommended for indoor installations in dry areas, especially where a good ground connection with a city water-piping system is available. Like BX cable, Greenfield is not intended for outdoor or underground installation. It may be used for either exposed or concealed work. These considerations serve as a guide for planning.

Rigid conduit (Figure 5-13) is comparable to water pipe; it is galvanized, and the inner surface is smooth to facilitate pulling wires through. Galvanized conduit can be used indoors or outdoors; black enamel conduit can be installed indoors only. The NEC prohibits rigid conduit through or beneath cinder fill in damp locations unless the conduit is buried at least 18 inches under the fill. Alternatively, rigid conduit may be enclosed by concrete 2 inches in thickness or more.

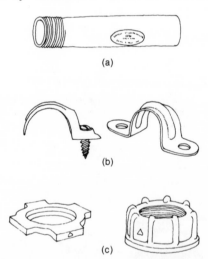

(a)

(b)

(c)

FIGURE 5-13   Rigid conduit and fittings   (a) Appearance of rigid conduit   (b) Straps used to support conduit   (c) Conduit-to-box connector

Most installations employ 1/2-inch rigid conduit; it is supplied in 10-foot lengths. Rigid conduit is threaded like water pipe and uses the same general kinds of fittings. In addition, numerous special conduit fittings are available, as illustrated in Figure 5-14. These are called condulet or unilet fittings. Note that conduit elbows with removable covers are used for pulling wires,

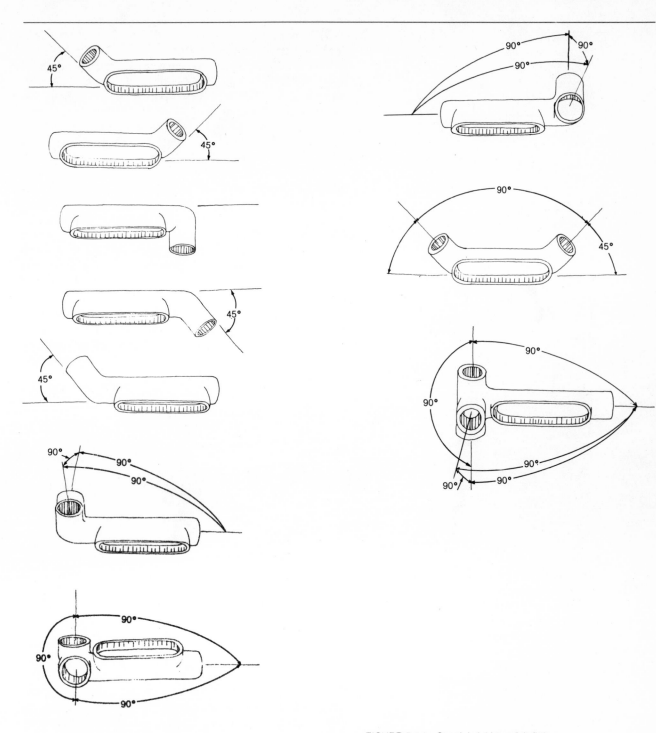

FIGURE 5-14   Special rigid-conduit fittings

as shown in Figure 5-15(a) and (b). Also shown is a conduit fitting used in securing the conduit to a switch box; see Figure 5-15(c). Although rigid conduit is more difficult to bend and install than thin-wall conduit, it makes a rugged installation and is sometimes required by local codes.

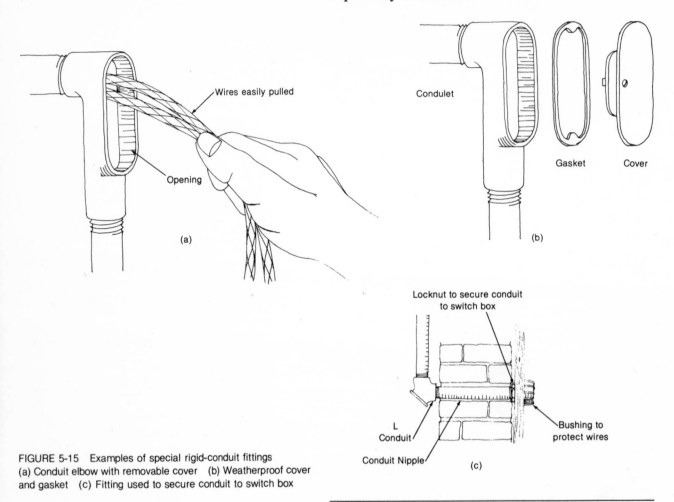

FIGURE 5-15   Examples of special rigid-conduit fittings
(a) Conduit elbow with removable cover   (b) Weatherproof cover and gasket   (c) Fitting used to secure conduit to switch box

SWITCH AND OUTLET BOXES

Switch and outlet boxes have been illustrated previously. There are various kinds of boxes, some of which are intended primarily for new work, with others intended for old work. Some boxes are made of steel, and others are made of plastic or porcelain. Figure 5-16 illustrates the appearance of a typical metal box with knockouts for cable attachment. This type of box can be used for either a switch or a receptacle outlet. When planning a wiring layout and making a list of material, remember that metal boxes can be ganged, if desired, as shown in Figure 5-17. It is only necessary to remove the wall end of each box, fit the boxes together, and tighten with screws. Note that plastic or porcelain boxes cannot be ganged. Although metal boxes are generally installed, plastic boxes are gaining favor for use with NM cable. Plastic or porcelain boxes must be installed with greater care than metal boxes, as they are easily cracked or broken. (See Figure 5-18.)

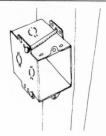

FIGURE 5-16   Metal box with knockouts for cable attachment

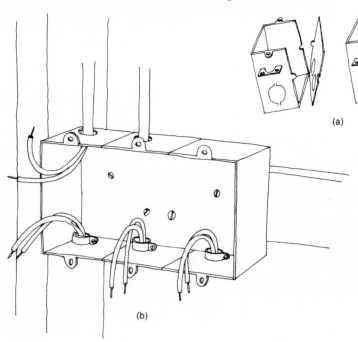

(a)

(b)

FIGURE 5-17   Ganging metal boxes   (a) Assembling a ganged box   (b) Appearance of three ganged boxes

Note that some boxes have internal cable clamps; these serve the same purpose as separate cable-to-box connectors. As an illustration, Figure 5-19 shows a box with internal cable clamps; this is a "square" box which is equivalent to a pair of ganged conventional

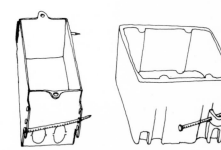

FIGURE 5-18   Two basic types of boxes   (a) Metal type (b) Plastic type

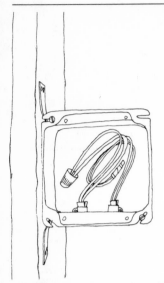

FIGURE 5-19 ''Square'' box with internal cable clamps and mounting bracket

boxes. Note also that the box in Figure 5-19 is provided with a mounting bracket (compare this with Figure 5-16). A mounting bracket speeds up installation. In old work, beveled corner boxes, such as the one shown in Figure 5-20(a), provide easier installation. Figure 5-20(b) shows a metal box support, which helps secure conventional boxes in old work. Octagonal

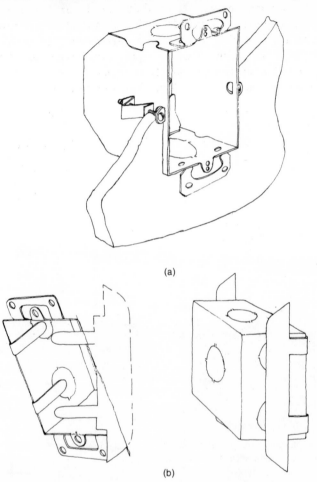

(a)

(b)

FIGURE 5-20 Typical boxes installed in old work (a) Beveled corner box, with grip-tite clamps (b) Metal support inserted with conventional box

boxes (Figure 5-21) are often used for mounting light fixtures, although boxes of other shapes are also installed for this purpose. Figure 5-22 depicts fixtures and mounting hardware for various boxes, with a view of the round box, which is often used for ceiling fixtures.

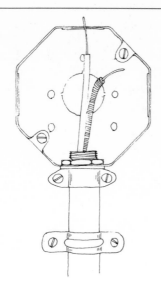

FIGURE 5-21   Octagonal box

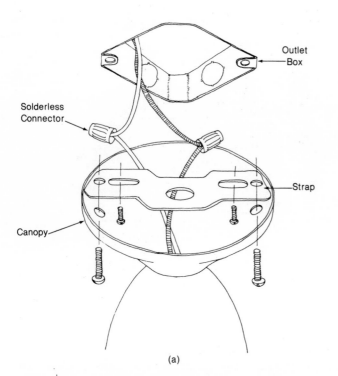

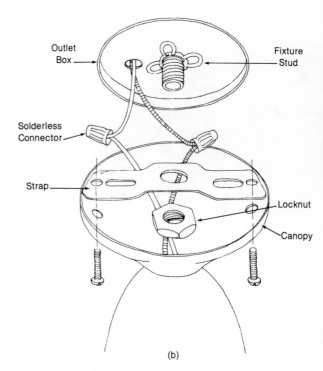

FIGURE 5-22   Outlet boxes and fixture mountings   (a) Ceiling
fixture with studs   (b) Ceiling fixture with fixture stud

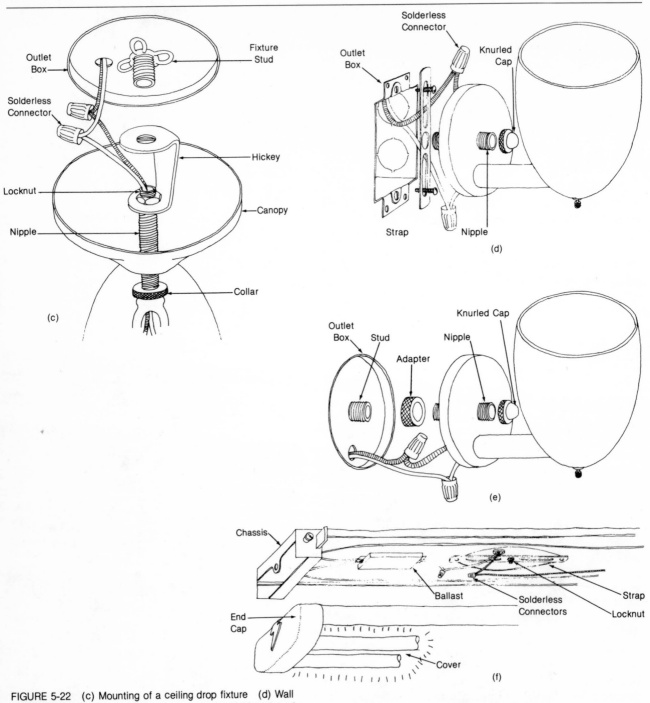

FIGURE 5-22   (c) Mounting of a ceiling drop fixture   (d) Wall
fixture with screws   (e) Wall fixture with stud   (f) Mounting of a
fluorescent fixture

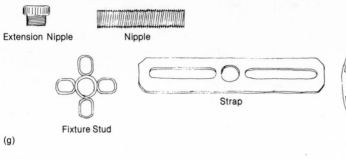

Extension Nipple   Nipple

Fixture Stud   Strap

(g)

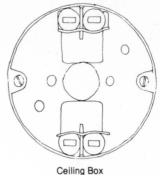

Ceiling Box

FIGURE 5-22   (g) Round box and ceiling fixture parts

A wiring list of material should also include the necessary box hangers (Figure 5-23) for ceiling fixtures, as well as plaster rings if required. When conduit is installed, the box must be mounted about a 1/2 inch behind the plaster surface. In turn, plaster rings or some variety of raised cover must be utilized to bring the assembly flush with the plaster surface.

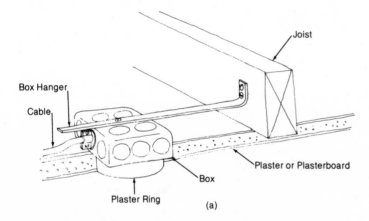

Joist

Box Hanger

Cable

Plaster or Plasterboard

Box

Plaster Ring

(a)

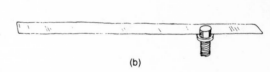

(b)

FIGURE 5-23   Box hanger provides support between joists or over ceiling   (a) Box hanger used in new work   (b) Sliding-type hanger used in old work

Cable connections must be enclosed in a box, as noted previously; conventional boxes or octagonal boxes are generally used for junction boxes. A plain metal plate is fastened over a box to make it a complete junction box. The total number of boxes required in a wiring system may depend upon whether copper wire or aluminum wire is used. That is, the NEC specifies the maximum number of wires that may be inserted into one box. For example, an octagonal box 3¼ inches across and 1½ inches deep can be used with a maximum of four No. 12 copper wires. If aluminum wire is used, a maximum of three conductors must be observed.

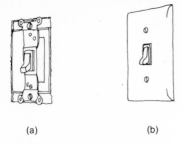

(a)   (b)

FIGURE 5-24   Single-pole double-throw toggle switch and plate
(a) Switch installed in box   (b) Switch plate installed

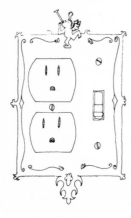

FIGURE 5-25   Decorative switch plate

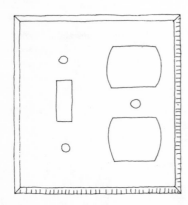

FIGURE 5-27   Plate for square box enclosing toggle switch and
duplex receptacle

SWITCHES AND RECEPTACLE OUTLETS

Most wiring installations employ toggle switches, as illustrated in Figure 5-24, and covered with a plastic plate. Silent switches have mercury contacts instead of mechanical members. Note that decorative switch plates are available, (Figure 5-25) fabricated from metal, hardwood, or porcelain, with various artistic designs. Three-way or four-way switches are installed in boxes in the same way as a single-pole double-throw (SPDT) switch. Sometimes, a lock-type switch is preferred, which can be turned on or off only by using a special key. Figure 5-26 shows a duplex receptacle mounted in the same kind of box. This is a grounding-type receptacle that accommodates a three-blade plug. However, the old-style two-blade plug will also fit into a grounding-type receptacle. Note that local codes may require grounding-type receptacles because of the protection that they provide from electrical shock if an appliance becomes defective.

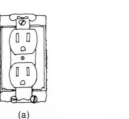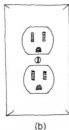

(a)   (b)

FIGURE 5-26   Duplex grounding-type receptacle and plate
(a) Receptacle installed in box   (b) Outlet plate installed

If a square box, or a ganged pair of single boxes is installed, an SPDT switch and a duplex receptacle can be installed in the same box and covered with a plate, as is shown in Figure 5-27. Again, if interchangeable devices like those shown in Figure 5-28 are used, a toggle switch, single receptacle, and pilot light may be mounted in a single box. Other combinations of devices may be chosen, if desired, such as two SPDT switches and a single receptacle. When planning switching arrangements, note that you can wire a duplex switch (Figure 5-29) to operate in a single branch circuit, or in

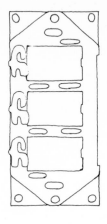
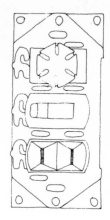

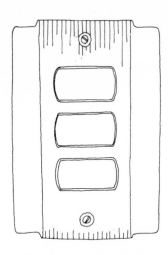
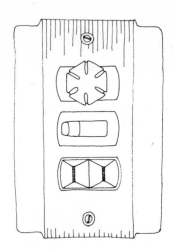

FIGURE 5-28 Three interchangeable devices that can be assembled on a skeleton strap

two branch circuits. Clock outlets (Figure 5-30) have recessed plates, so that an electric clock can hang flush with the wall surface.

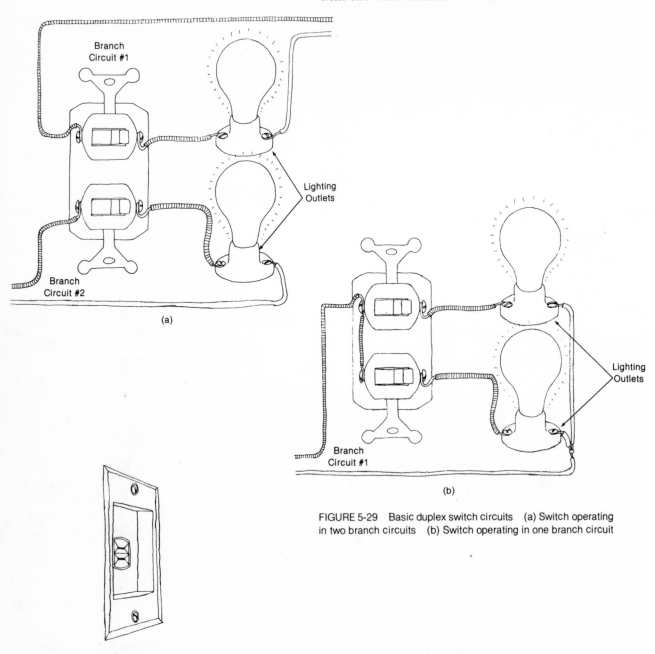

(a)

(b)

FIGURE 5-29   Basic duplex switch circuits   (a) Switch operating in two branch circuits   (b) Switch operating in one branch circuit

FIGURE 5-30   Clock outlet with recessed plate

Outdoor boxes, receptacles, and switches (Figure 5-31) are weatherproofed to prevent entry of moisture. Receptacles are covered with spring-loaded caps; switches may also be covered with spring-loaded caps, or may be designed with rotary construction. Boxes are provided with four 1/2-inch watertight openings for cable installation. Note carefully that outdoor receptacles in various locations must be combined with special built-in relays called ground-fault circuit interrupters (GFCI). Outlets in swimming-pool areas must be protected with GFCIs, and the NEC recommends their use on all outdoor circuits. A GFCI contains a sensitive relay that opens the circuit in 0.025 second if a ground current of 5 milliamperes or more should occur. Thus, a GFCI supplements the function of the third-conductor ground wire. A GFCI provides necessary protection because, in case an appliance develops a short-circuit to the frame, such a large fault current can flow that the voltage drop across the ground wire to ground is sufficient to kill a person. Not only does a GFCI operate on a very small current, but it also operates far more rapidly than a conventional circuit breaker.

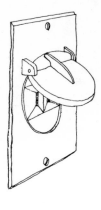  

FIGURE 5-31   Weatherproof receptacle, switch, and box

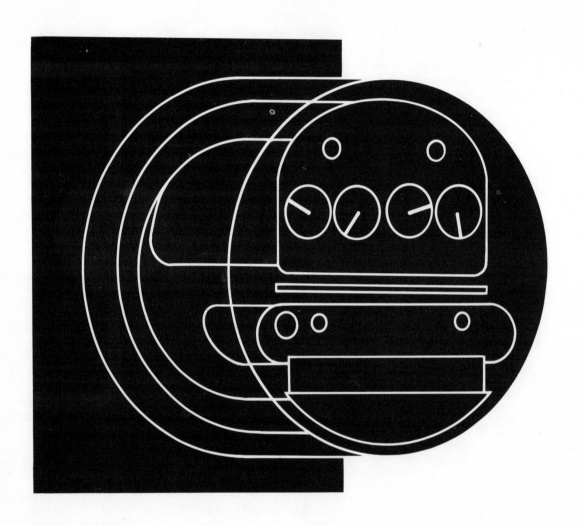

HOW TO LAY OUT A SERVICE ENTRANCE 6

GENERAL CONSIDERATIONS

In planning a new residential electric service installation, it is important to provide your local public-utility office with accurate load data. If local ordinances require permits and inspection, you must obtain these before the public utility can connect the service wires. There is a trend to underground service in areas that have an existing underground distribution system. The underground service lateral (extending in a horizontal direction) is installed, owned, and maintained by the public utility from the utility's distribution line to the kwh meter. (See Figure 6-1.) The owner of the residence is required to provide or make arrangements and pay for the trenching and backfill in accordance with the public utility specifications. If the total length of the service line is 100 feet or less, the conductors are provided at the utility's expense. On the other hand, if the service line exceeds 100 feet, the owner of the residence is charged for the material cost and any required conductors in excess of 100 feet.

Electric service conductors are ordinarily installed, along with gas, telephone, and any other wire service

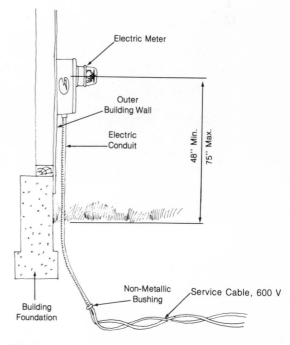

FIGURE 6-1   Residential service installations (general arrangements)

facility, in a single trench of adequate size to accommodate those facilities in joint occupancy. (See Figure 6-2.) When such a joint trench is utilized, the residence owner may excavate it himself to utility specifications, or the utility will do the trenching on request. If done by the utility, the cost of the trench is allocated equally among the different utility facilities in joint occupancy, and the residence owner is billed for the electric share plus any other utility's share that he may be required to provide. The public utility must be consulted while the building is in the planning stage in order to determine the most satisfactory common location of the gas and electric meters. Preferred meter location is depicted in Figure 6-1. Meters are generally located within 36 inches of the wall nearest to the street or easement where the utility's distribution facilities are located.

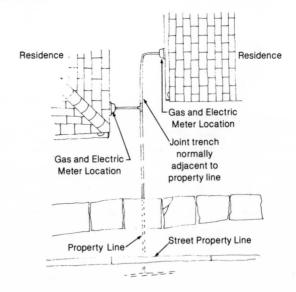

FIGURE 6-2 Location of meters, service lines, and joint trench

OVERHEAD LINES

If the building is located in an area served from an overhead system and you prefer to have your service installed underground, or if the city or county requires underground service by ordinance, service may be provided from an underground riser installed on an existing pole. In this case, in addition to the foregoing requirements, the residence owner must pay the material costs of the pole riser facility, any required conduit in the public right of way, and any conductor in excess of 100 feet measured from the top of the riser. In areas served from overhead lines, an overhead service drop will be installed at utility expense from the distribution line to the point of attachment on the residence. This point of attachment is generally on the building wall facing the nearest utility line, or on a periscope mounted on the roof, as illustrated in Figure 6-3. The height to the point of attachment must be sufficient

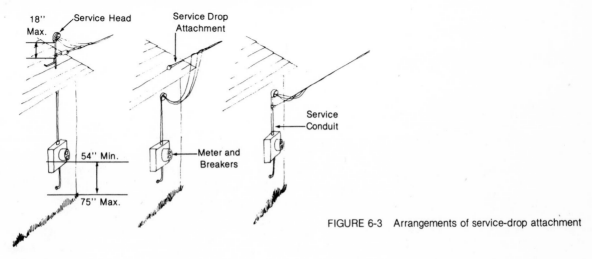

FIGURE 6-3   Arrangements of service-drop attachment

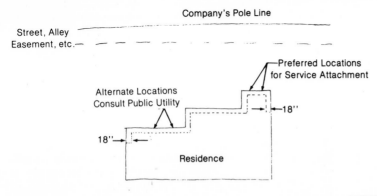

FIGURE 6-4   Location of overhead service-drop attachment

to provide the minimum service-drop clearance required by the public utilities commission. These clearances are shown in Figures 6-4 and 6-5.

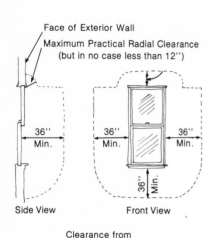

Clearance from
Windows, Doors, Exits, etc.

FIGURE 6-5   Clearance requirements

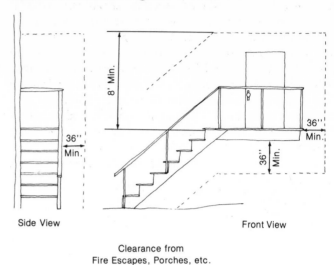

Clearance from
Fire Escapes, Porches, etc.

In case a normal location is unsuitable, you should consult the public utility. The residence owner must provide a substantial support for the overhead service drop attachment. An approved rain-tight service head must be furnished and installed by the residence owner on the periscope or adjacent to the point of attachment of the service drop. The service-entrance wire should extend 18 inches outside of the service head in order to provide sufficient length for drip loops and attachment to the utility's overhead service drop. The utility furnishes and installs connectors for splicing the overhead service-drop wires to the service-entrance wires. The service must have a grounded neutral conductor that is securely connected to the neutral terminal of the meter socket, and extended to the neutral terminal of the service-entrance switch. In the case of an underground service, the utility will connect the service neutral to the neutral terminal of the meter socket, and it is necessary for the residence owner to provide the connection from the socket to the service switch. The neutral terminal of the service entrance must be grounded in accordance with applicable local regulations. In those areas where there are no applicable

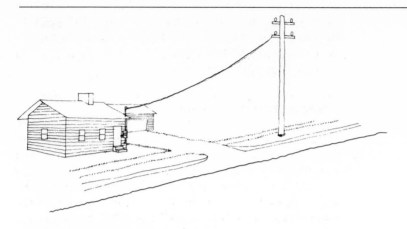

VERTICAL CLEARANCE ABOVE RAILS
(a) Crossing above RR tracks without overhead trolley wire . . . . 25 ft.
(b) Crossing above RR tracks with overhead trolley wire:
    Above rails where freight cars are transported . . . . 26 ft.
    Above rails where freight cars are not transported . . . . 23 ft.
In each case the service frops must clear trolley wires by not less than 4 ft.

VERTICAL CLEARANCE ABOVE GROUND
(a) Crossing public thoroughfares:
    Above center portion, between points 12 ft. horizontally
    from curbs . . . . 18 ft.
    At curb line . . . . 16 ft.
    Where there are no curbs, the curb line shall be taken as the outer
    possible limit of vehicular traffic.
(b) On premises accessible to agricultural equipment
    over private roads and other areas . . . . 16 ft.
    Public Utility Commission permits a clearance of 15 ft. Maintain
    16 ft. if possible.
(c) On residential premises (not accessible to agricultural equip.)
    Over private driveways or other areas accessible to vehicles . . . . 12 ft.
        If 12 ft. clearance requires a structure on the building
        served, 0-300 volt service drops may have a clearance of . . . . 10 ft.
    Over areas accessible to pedestrians only . . . . 10 ft.
        If 10 ft. clearance requires a structure on the building
        served, 0-300 volt service drops may have a clearance of . . . 8½ ft.
    If clearance also requires a structure on the building served, then
    the greater minimum clearance must be maintained.

VERTICAL CLEARANCE ABOVE STRUCTURES
Includes buildings, bridges and other structures on which men can
walk and applies whether wires are attached or unattached.

CLEARANCE ABOVE BUILDINGS (FT.)

TYPE OF ROOF	Above Bldg. served	Above other Bldg. on premises served	Above Bldg. on other premises
Metal, 3/8 pitch or less*	1	8**	9
Metal, more than 3/8 pitch	1	2	8
Nonmetallic, 3/8 pitch or less	No limit***	2	8
Nonmetallic, more than 3/8 pitch	No limit***	2	2

  *3/8 pitch is a slope of 3 ft. vertical in 8 ft. horizontal or 30° from horizontal
  **May be reduced to 2 ft. if roof is non-walkable overhang or patio cover
***Maintain greatest practicable clearance

CLEARANCE ABOVE SWIMMING POOLS
Public Utility Commission requires greater clearances for service drops over
or near swimming pools; consult them for requirements.

FIGURE 6-6   GROUND CLEARANCE FOR SERVICE DROP
Locate service drop attachment to building according to the
minimum vertical clearances given. Allow for sag of conductors
at a temperature of 60° F., no wind

local ordinances, the provisions of the NEC must be followed. Note that a grounding connection must not be made to a gas pipe. Of course, plastic pipe does not provide a ground connection.

TERMINATION FACILITIES

With regard to termination or service-entrance facilities, the residence owner is required to furnish and completely wire a 4-terminal meter socket (5-terminal for 120/240-volt service) without circuit-closing devices, as shown in Figure 6-7. If the capacity of the main service switch does not exceed 125 amperes, any UL-approved socket may be used, provided that it has beryllium-copper alloy (or equivalent) jaws, approved terminals for use with aluminum as well as copper conductors, and, in the case of underground service, an insulated two-way neutral terminal lug for connecting the aluminum service cable neutral wire to the neutral conductor of the residence wiring system. In case the capacity of the residential service switch or breaker is 126 to 200 amperes, only 200-ampere sockets approved by the public utility may be installed.

The meter socket must be mounted so that the socket jaws are vertical and plumb, regardless of whether the conduit or cable enters the socket vertical-

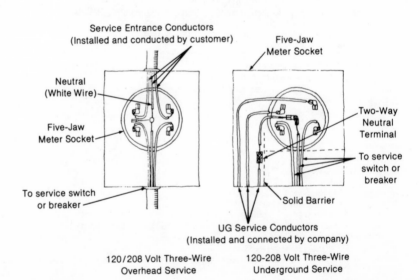

FIGURE 6-7 Connections for meter socket

ly or horizontally. The meter socket must be located on the source side of the service switch. The sequence of equipment is: service-entrance wiring, meter, switch, and then fuse or circuit breaker. The main service switch or breaker must be at a readily accessible location near the entrance of the conductors, either inside or outside the building wall. If the service-entrance switch is installed in a location exposed to the weather, it must be of waterproof design or protected by a weatherproof enclosure. Suitable standing space (at least 3 feet) must be maintained in front of the meter to allow installation, testing, and reading. Eight inches of clear space is normally required below the center line of the socket, but this clearance may be reduced to 6 inches as shown in Figure 6-8 if the obstruction is a service switch.

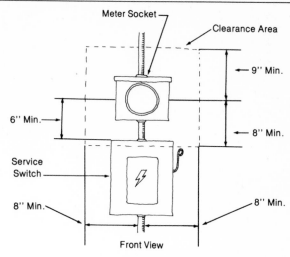

FIGURE 6-8   Meter socket clearance requirements

CONDUCTOR SUPPORTS

When conductors pass over buildings, the NEC requires that supports shall be used where it is practical, as exemplified in Figure 6-9. Supports are also used where necessary to raise the conductors over the building to provide required clearance. A support must be substantially constructed, securely installed, and provided with approved insulators. Occasionally, service conductors must be installed horizontally along the side of a building, as depicted in Figure 6-10. In such a case, the conductors must be at least 8 feet above

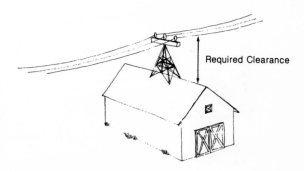

FIGURE 6-9   Service conductor support

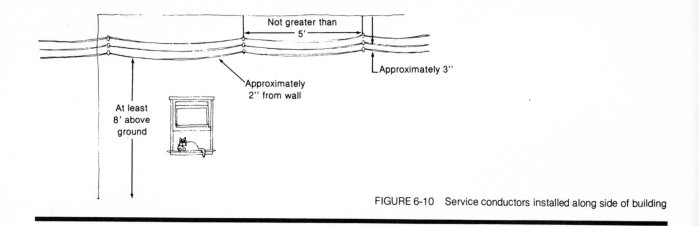

FIGURE 6-10   Service conductors installed along side of building

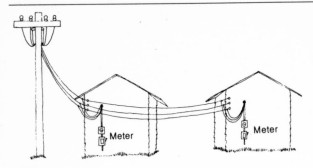

FIGURE 6-11 Service conductors for two buildings are generally installed outside of both buildings

ground, and supported by approved insulators at intervals not greater than 5 feet. The insulators must be of a type that provides at least 2 inches of clearance between the conductors and the wall.

If two buildings are under the same ownership, the service entrace is generally installed as shown in Figure 6-11. In other words, the NEC seldom permits a service to run through one building to another building.

YARD POLE SERVICE ENTRANCE

Yard pole service entrances are installed in some situations. For example, if electricity is required at a work

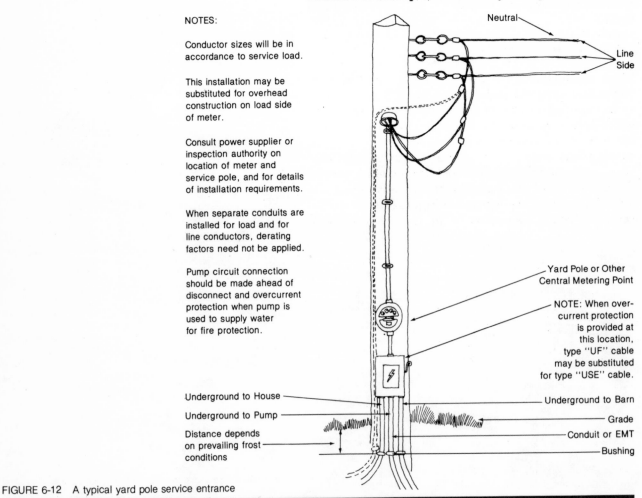

NOTES:

Conductor sizes will be in accordance to service load.

This installation may be substituted for overhead construction on load side of meter.

Consult power supplier or inspection authority on location of meter and service pole, and for details of installation requirements.

When separate conduits are installed for load and for line conductors, derating factors need not be applied.

Pump circuit connection should be made ahead of disconnect and overcurrent protection when pump is used to supply water for fire protection.

Neutral

Line Side

Yard Pole or Other Central Metering Point

NOTE: When over-current protection is provided at this location, type "UF" cable may be substituted for type "USE" cable.

Underground to House

Underground to Pump

Distance depends on prevailing frost conditions

Underground to Barn

Grade

Conduit or EMT

Bushing

FIGURE 6-12 A typical yard pole service entrance

site that has no suitable building structure to support a service entrance, a yard pole is generally utilized. Yard pole service entrances are often installed on farms, as depicted in Figure 6-12. In this example, the service drop is supported at the top of the yard pole, and the service entrance feeds underground conductors running to a farmhouse and various outbuildings. Figure 6-13 shows the materials that are required for a typical yard pole installation. In this example, the service entrance feeds overhead conductors running to the various buildings. In any case, a yard pole service entrance should be laid out with the assistance and advice of the local public utility or inspection authority.

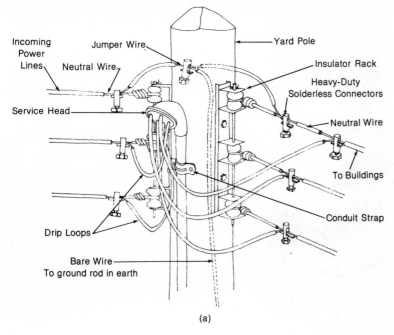

(a)

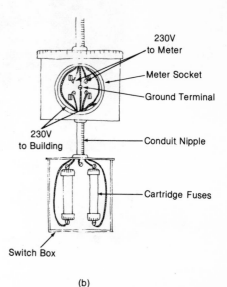

(b)

FIGURE 6-13   Components of a yard pole service entrance
(a) Assembly at top of pole   (b) Switch box and meter assembly

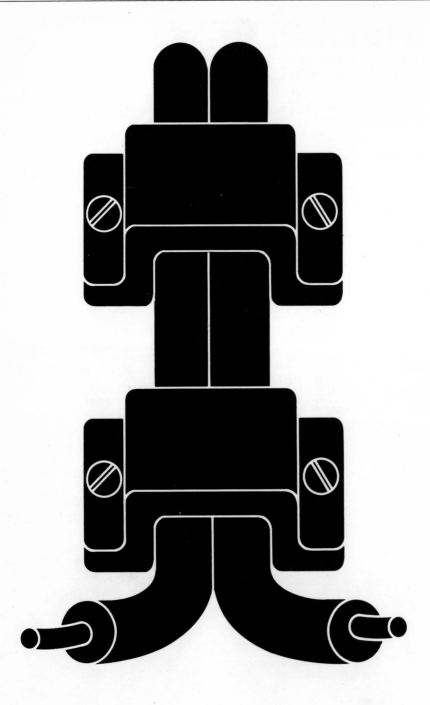

HOW TO INSTALL ELECTRICAL WIRING 7

GENERAL CONSIDERATIONS

Residential wiring is installed with nonmetallic cable, thin-wall conduit, rigid conduit, or armored cable. Nonmetallic cable requires stripping at the end before installation. Strip the jacket so that at least 8 inches of insulated wire is provided for connection purposes. As shown in Figure 7-1, fasten a connector to the jacket past the stripped ends, and insert the connector into the knockout hole in the box. Then screw the locknut up tightly on the connector from inside the box. You can remove knockouts from boxes by a sharp blow with a screwdriver, by prying out (in case a slot is provided), or by means of a pair of pliers (if located near the edge of the box). Nonmetallic cable should be strapped every 3 feet along supporting surfaces such as studs, joists, ceilings, or walls. When the cable is run horizontally, holes are bored through studs, as illustrated in Figure 7-2.

Where the cable is installed crosswise of joists, holes may be bored through the joists, or the cable may be strapped to a running board, as depicted in Figure 7-3. A running board usually consists of 1"×2" wood, and

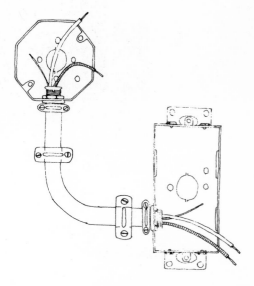

FIGURE 7-1   Installing nonmetallic cable through knockouts in boxes

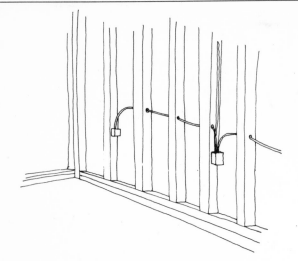

FIGURE 7-2   Studs are bored for horizontal installation of cable

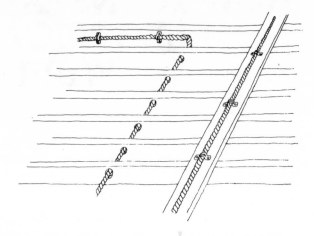

FIGURE 7-3   Cable installed on running board and cable passing through holes in the joists

is used when the cable is installed below the joists. On the other hand, if the cable is installed above the joists, it must be run between a pair of guard strips, as shown in Figure 7-4. The guard strips have a thickness at least equal to that of the cable, and prevent damage to the cable from objects placed in the attic, or from persons walking about. In concealed work, cable should

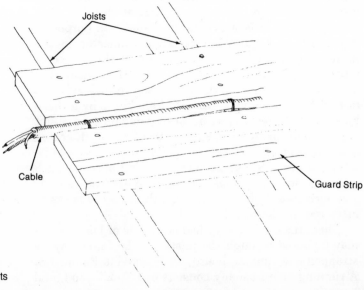

FIGURE 7-4   Cable protected by guard strips on top of joists

be strapped every 4½ feet and within 12 inches of all outlets and switches. However, in old work, straps are not used in concealed runs; exposed runs in old work must be strapped.

INSTALLING GROUND WIRE

The ground wire (bare wire) in nonmetallic cable must be connected to the box by means of a clip or a 10-32 machine screw, as shown in Figure 7-5. By providing a continuous system ground which is connected to the boxes, the possibility of shock is virtually eliminated. Note that, although the white neutral wire is grounded, the neutral wire is never connected to a box. When plastic or porcelain boxes are installed, they are not grounded, of course. This type of box inherently eliminates the possibility of shock. When wiring a grounding receptacle (Figure 7-6), connect the ground wire in the nonmetallic cable to the Gr terminal of the receptacle. If the receptacle is installed in a metal box, the ground wire connects both to the Gr terminal and to the box.

INSTALLATION OF ARMORED CABLE

Armored cable is installed somewhat in the same manner as nonmetallic cable. However, it must be used with metal boxes only. A fine-toothed hacksaw is used to remove the spiral-wound steel armor from the cable. Figure 7-7 shows how to hold the saw at right angles to the spirals. Carefully cut through one section of the armor, so that the saw goes through the armor but does not cut into the insulation of the wires. Next, grasp the cable on each side of the cut, and twist off the short end. The cable must then be prepared for insertion of a fiber bushing at the cut end. To provide room for the bushing, unwrap the paper from the around the wires up to a few turns under the armor. Then pull the wrapping sharply, so that it will tear off inside the armor. Then insert the protective fiber bushing as depicted in Figure 7-8. Note that if the armored cable has a bare bonding wire (Figure 7-9), you should bend it back before inserting the bushing.

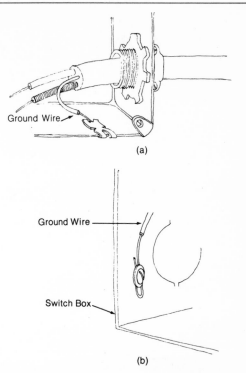

(a)

(b)

FIGURE 7-5   Connection of bare ground wire to boxes   (a) With grounding clip   (b) With grounding screw

FIGURE 7-6   Ground wire of nonmetallic cable connects to GR terminal on receptacle

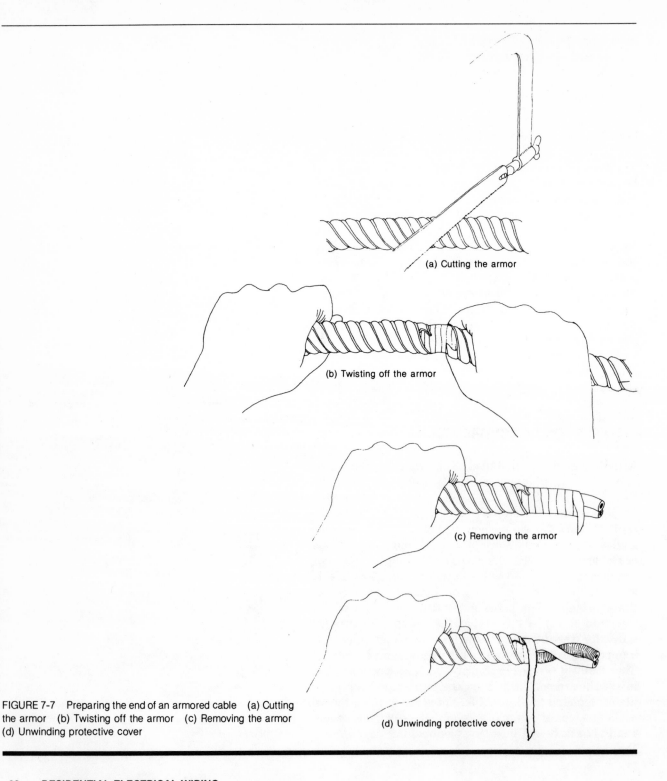

(a) Cutting the armor

(b) Twisting off the armor

(c) Removing the armor

(d) Unwinding protective cover

FIGURE 7-7   Preparing the end of an armored cable   (a) Cutting the armor   (b) Twisting off the armor   (c) Removing the armor   (d) Unwinding protective cover

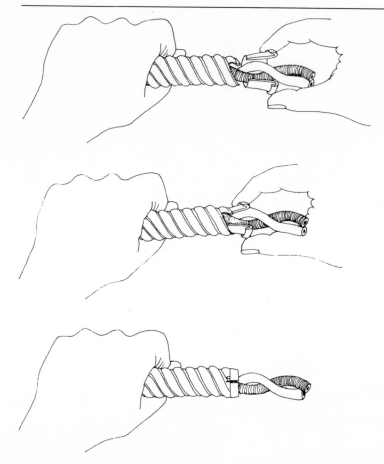

FIGURE 7-8   Inserting the fiber bushing

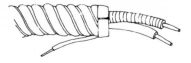

FIGURE 7-9   Armored cable with bonding wire

In case the outlet box has a cable clamp, secure the armored cable by tightening the clamp over it. On the other hand, if the box does not have a cable clamp, use a cable connector. Slip the connector over the cable after the fiber bushing, and tighten the holding screw as illustrated in Figure 7-10. Punch out a knockout slug in the box, and insert the connector into the hole.

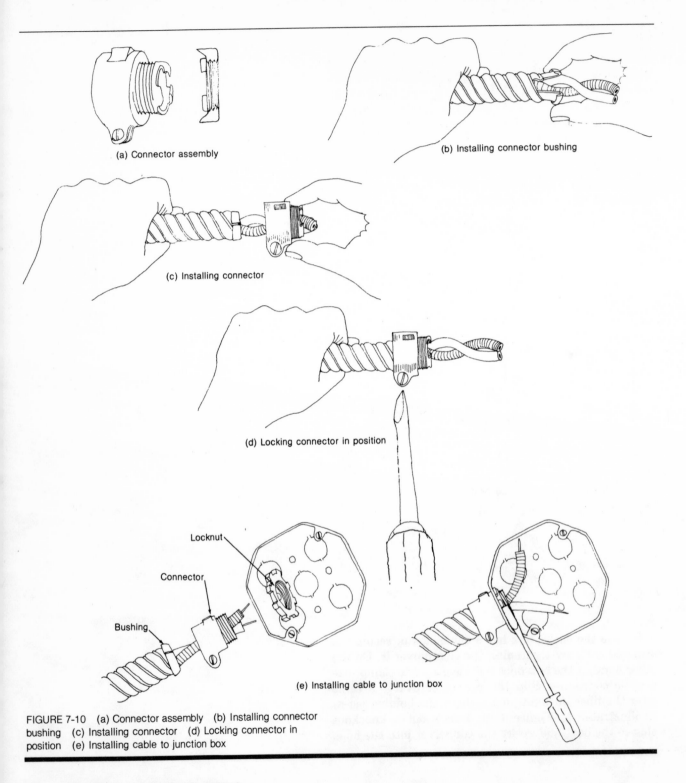

(a) Connector assembly

(b) Installing connector bushing

(c) Installing connector

(d) Locking connector in position

Locknut

Connector

Bushing

(e) Installing cable to junction box

FIGURE 7-10   (a) Connector assembly   (b) Installing connector
bushing   (c) Installing connector   (d) Locking connector in
position   (e) Installing cable to junction box

Then turn up the locknut tightly to secure the connector to the box. If the cable has a bonding wire, this wire is connected to the box, as explained previously for NM cable. When a grounding receptacle is installed, the bonding wire is also connected to the Gr terminal on the receptacle. Armored cable is supported by straps or staples every 4½ feet, and also within 6 to 12 inches of every outlet or switch box. However, on concealed runs in old work, armored cable is left unsupported. In some situations, a 90° connector may be utilized at a box, as shown in Figure 7-11.

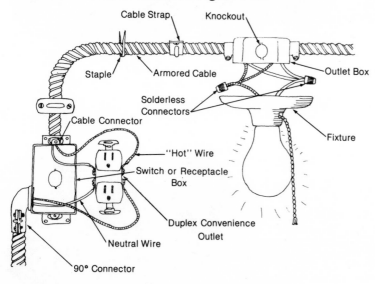

FIGURE 7-11   An armored-cable installation with a 90° connector

INSTALLATION OF THIN-WALL AND RIGID CONDUIT

Thin-wall conduit, also called electrical metallic tubing (EMT), is installed in much the same way as rigid conduit, except that different types of connectors and bushings are utilized. Like BX, thin-wall conduit is used only with metal boxes. Since conduit provides a good continuous ground for a system, a bonding wire is not used. Both thin-wall and rigid conduit must be bent, in routing an installation, as illustrated in Figure 7-12. A conduit bender (hickey) is used for this purpose, as also depicted in Figure 7-12. Bends must not

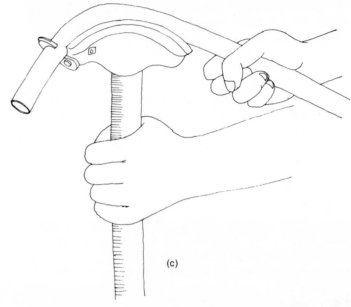

be so sharp that the conduit is kinked or collapsed; sharp bends also make it difficult to pull the cable or wires through the completed run. It is good practice to make not more than four 90° bends on a single conduit run.

Both thin-wall and rigid conduit are cut with a hacksaw. After cutting, it is important to remove the sharp edges with a reamer and a file. If sharp edges or burrs

(a)

30'' Pipe Handle

(b)

(c)

(d)

FIGURE 7-12   Installation of conduit   (a) Example of three 90° bends   (b) Appearance of a hickey   (c) Insertion of conduit into hickey   (d) Conduit-bending process

are permitted to remain, insulation may be damaged and it may be difficult or impossible to slide fittings over the conduit. Figure 7-13 shows how to join sections of thin-wall conduit with compression couplings, and how to secure the end of a run to a box with a compression connector. In the case of rigid conduit, thread it in the same manner as water pipe, and install it with threaded connectors and couplings. If a run of rigid conduit is to be continued with thin-wall conduit, use a threaded compression coupling to mate the two kinds of conduit.

Conduit is anchored to supporting surfaces with a pipe strap every 6 feet on exposed runs and every 10 feet on concealed runs. After the conduit is mounted in place and secured to switch and outlet boxes, pull the conductors through by means of a snake or equivalent line. A snake is also called a fish tape. Type TW wires are generally pulled through a conduit run. The wires are color-coded as previously described. Note that only continuous wires can be pulled through a run of conduit. As in any electrical wiring arrangement, splices and connections are permitted only inside of switch, outlet, or junction boxes. Figure 7-14 shows how to join a fish tape to a pair of wires, after pushing the fish tape through the conduit. If a black wire, white wire, and red wire are to be installed, pull all three wires together through the conduit. Approximately 8 inches of free end should be left hanging when the wires are pulled out of a box, to provide ample length for subsequent connections.

Greenfield (flexible conduit) is similar in appearance to armored cable, but it has a larger diameter to facilitate pulling wires through. As noted previously, Greenfield is installed chiefly in old work; it can often be pulled behind walls through holes cut for boxes, thereby minimizing or eliminating removal of plaster and lath or wallboard. It is secured to boxes in the same manner as explained previously for armored cable. Since Greenfield is not supplied with a bonding wire, it is usually necessary to pull a grounding wire along with the black and white, or black and white and red wires. It shows foresight to consult with your local public utility or inspection authority. Note that in many localities, all new work must be installed with thin-wall or rigid conduit, and both armored cable and Greenfield are forbidden.

Conduit may be run through notches in studs, as shown in Figure 7-15, if notching is preferred to drilling. Note that steel box supports can be used to mount receptacle boxes between studs; they are also used to mount one or more switch boxes in any chosen position. If preferred, thin-wall conduit can be run across subfloors as illustrated in Figure 7-16, instead of notching or drilling studs. An advantage of this method is that it does not weaken the studs; it also involves less work, because the only operation required is to make a small notch in the plate. In any installation, mount boxes carefully so that their edges will be even with the finished wall surface. Use plaster rings, if necessary, as explained previously. Do not mount a box closer than 2 inches from a door frame.

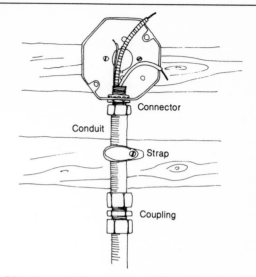

FIGURE 7-13   Installation features of thin-wall conduit

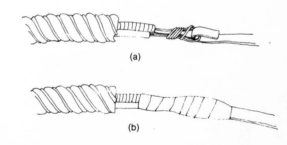

FIGURE 7-14   Preparations for pulling wires   (a) Joining of fish tape and wires   (b) Taping the joint makes pulling easier

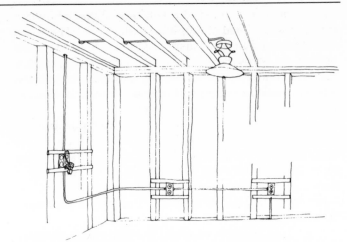

FIGURE 7-15   Notching studs for a run of conduit

When installing conduit for a ceiling outlet, as depicted in Figure 7-17, with a finished floor to be laid overhead, form a slot as shown. Drill two holes 6 inches apart in line with conduit, and then saw out a

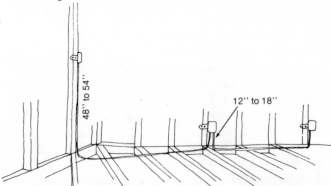

FIGURE 7-16   Conduit run across a subfloor

slot. A keyhole saw is suitable for this purpose. Unless a slot is made, the conduit would have to be bent with too small a radius. When nailing the hanger for the ceiling outlet in place, check the level of the outlet with respect to the joists and make certain that the edge of the outlet will be even with the finished ceiling. Mounting of fixtures on ceiling outlets was illustrated in Chapter 5.

INSTALLATION OF FUSED AND FUSELESS PANELS

When conduit is installed for a service entrance, it must make a good electrical connection to the service

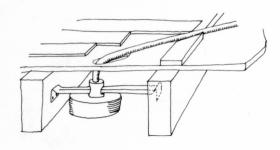

FIGURE 7-17   Conduit run to a ceiling outlet

enclosure, as shown in Figure 7-18. A ground wire is run from a cold water pipe and is connected to the service enclosure and to the neutral strip. A bare ground wire is generally installed. In case flexible conduit is utilized (Figure 7-19), it must be bonded with a heavy wire connected to approved clamps. A bonding conduit

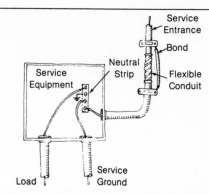

FIGURE 7-19   Bonding of flexible conduit for a service entrance

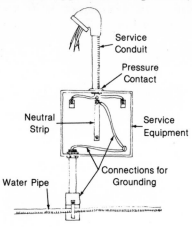

FIGURE 7-18   Installation of conduit for service entrance

connector must also be installed and connected to the neutral strip. Note that bonds must be connected by means of pressure contact, and must not be soldered. In general, No 4 ground wire is recommended for service entrances up to and including 200 amperes. In a rural installation, note that the ground wire does not go through the entrance switch, but is tapped off from the neutral overhead wire and run down to a ground rod or an underground water pipe. Figure 7-20 shows the differences between urban and rural installations. A ground rod may be an 8-foot length of 1/2-inch copper, or a galvanized iron or steel pipe at least 3/4 inch in diameter. The rod must be driven at least 1 foot below the grade level and at least 2 feet from any building.

OLD-WORK INSTALLATION PROCEDURES

The chief consideration in old work is determining the best locations of openings for fishing wires through walls, above ceilings, or under floors with minimum

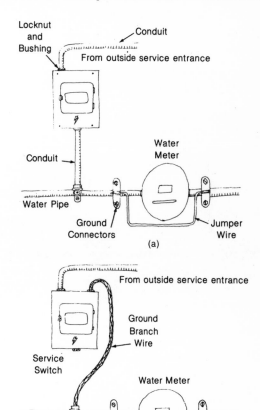

FIGURE 7-20   (a) Urban grounding arrangement   (b) Urban grounding arrangement using armored cable

difficulty and with least removal of existing structures. Therefore, a thorough evaluation and careful planning of each job is required. When there is more than one acceptable location for an outlet or switch, good planning dictates that the final decision should rest upon details of installation procedure. A typical cable installation from a ceiling outlet to a wall toggle switch is depicted in Figure 7-21. Note that at point

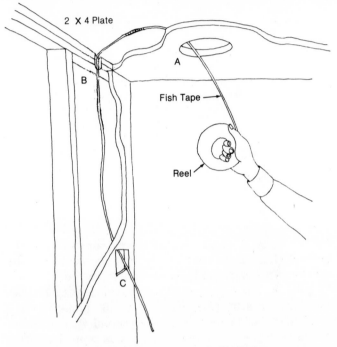

FIGURE 7-21   Ceiling outlet and wall switch installation (a) Hole cut for outlet in ceiling   (b) Opening cut in wall, notch chiseled in plate   (c) Hole cut in wall for switch

B, where the wall joins the ceiling, there is a 2" X 4" plate. To pass the plate, it is customary procedure to make a temporary opening in the wall. The 2 X 4 is then notched to provide passage for the cable, after which the opening is repaired. In the case of a plastered wall, patching plaster is applied to restore the surface appearance.

It is often necessary or helpful to install shallow outlet boxes in old work; a minimum depth of a 1/2 inch is permitted. Figure 7-22 shows the features of a shallow box. When there is no available space to work above the ceiling, installation is made from below, as

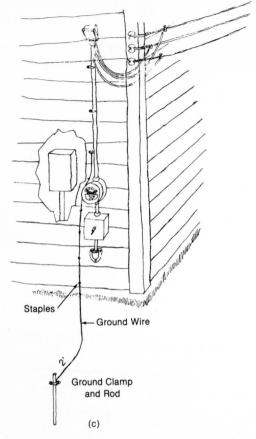

FIGURE 7-20   (c) Rural grounding arrangement with a ground rod

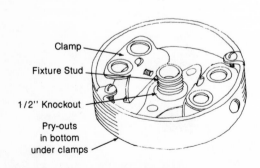

FIGURE 7-22   Features of a shallow box

shown in Figure 7-23. First, remove the plaster to make a hole the size of a shallow box, and cut the center lath to provide an opening. Next, insert an old-work hanger, with the locknut removed and the wire

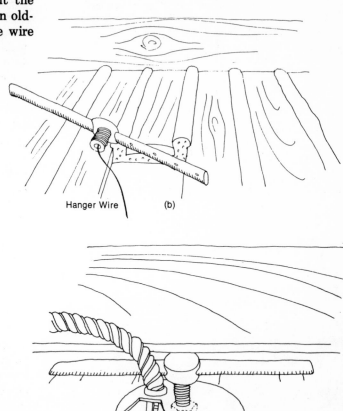

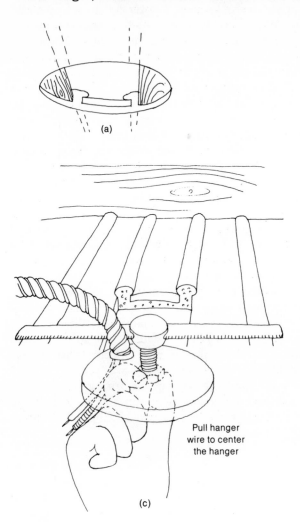

FIGURE 7-23   Installation of ceiling box   (a) Making the opening   (b) Inserting the hanger   (c) Connecting the cable   (d) The completed installation

passed through the threaded stud. By holding the stud in position with one hand and pulling on the wire with the other hand, the hanger will automatically be centered. Then secure the cable to the box and pull the wire from the hanger through the center knockout. Finally, secure the box to the hanger by screwing the locknut on the threaded stud.

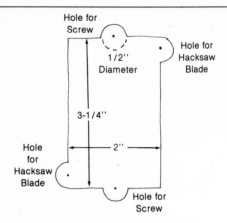

FIGURE 7-24   Size of opening for switch or outlet box

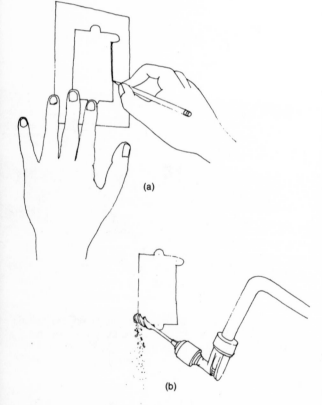

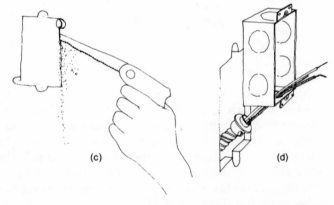

FIGURE 7-25   Installation of a switch or receptacle box   (a) Outlining the opening   (b) Drilling 1/2-inch holes   (c) Sawing out the opening   (d) Preparing box for insertion

All wall switches should be installed at the same height above the floor. This height may be in the range from 48 to 54 inches. Convenience outlets are installed from 12 to 18 inches above the floor. Convenience outlets over a kitchen countertop should be about 45 inches above the floor. As noted previously, special boxes are installed for switches and convenience outlets in old work. These boxes are inserted through suitable openings made in the wall. Figure 7-24 shows the size of opening that is required. Of course, the opening must be made clear of studs or other obstructions. A template should be made for outlining the required opening, as depicted in Figure 7-25(a). Then four 1/2-inch holes are drilled as shown. Next, a keyhole saw or a hacksaw blade with a taped handle is used to make the cutout. Note that it is helpful to use a hacksaw blade with the teeth backward, so that the cutting is accomplished by pulling on the blade. This minimizes chipping of plaster. Insert the box and attach the cable. Finally, secure the box in place; it is now ready for connection of the switch or receptacle and attachment of a cover plate.

Studs behind a wall can usually be located by rapping on the wall with your knuckles. A magnetic nail detector is also useful for this purpose. When sawing out an opening in a plaster wall, it is helpful to hold one hand against the plaster near the cut. This lessens the likelihood of chipping. Ready-mix plaster is used to repair accidental chipping. In wiring single-story houses, it may be practical to run the cable through the attic. Attic floor boards can be lifted, and holes drilled through joists to pass cable as in new work. The floor boards are then replaced. Lifting of finished floor boards is a tricky job and should not be attempted unless you have practical experience in this type of carpentry. It is sometimes necessary to run cable from a second floor to a first floor. When the second-floor partition is located directly over the first-floor partition, it is generally an easy procedure to run the cable from the upper floor by removing a baseboard and then boring a hole through the floor and 2 X 4 plate. After the cable is installed, the baseboard is replaced. To prepare a route and fish a cable from a ceiling outlet to a wall switch, proceed as shown in Figure 7-26.

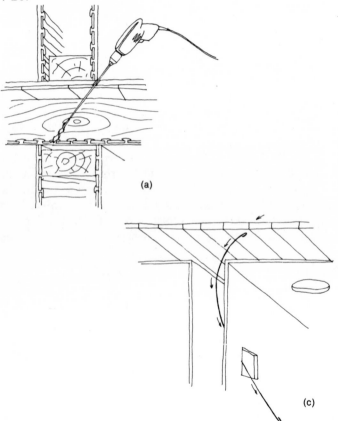

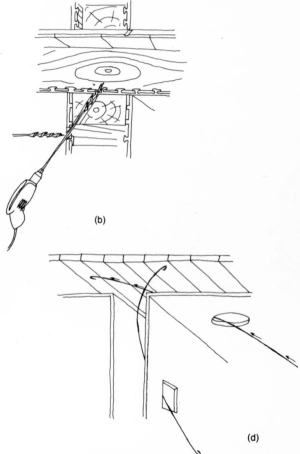

(a)

(b)

(c)

(d)

FIGURE 7-26 Running cable from upper floor to lower floor (a) Remove baseboard and drill diagonal hole (b) Drill horizontally to start, then drill diagonal hole upward (c) Work fish wire through baseboard hole and hook out through wall switch hole (d) Work another fish wire from outlet hole

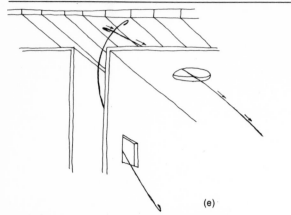

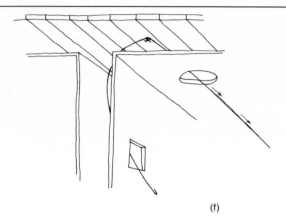

(e)

(f)

FIGURE 7-26   (e) Draw on both hooks together   (f) Pull wire through switch outlet hole to draw cable through

Cable from one outlet to another can often be run behind a baseboard as shown in Figure 7-27. Openings are made for the boxes, and the baseboard is removed. Cut small holes in the wall at points 3 and 4 directly below the boxes, as shown. Then notch a channel for the cable along the lath and plaster, or between a pair of laths. Cut a suitable length of cable to run between outlets 1 and 2 and place the cable in the notch way; fish cable up to the holes for the boxes. Finally, replace the section of baseboard. Alternatively, the cable can be run through the floor and across a basement, as illustrated in Figure 7-28. Drill holes upward from the basement into the partition between the walls. Fish the cable up to the holes for the boxes. Strap the cable across the basement every 3 feet to secure it to beams

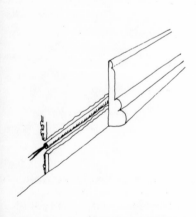

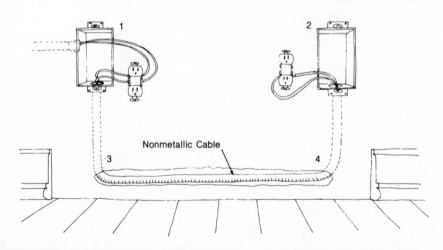

FIGURE 7-27   Cable run behind a baseboard

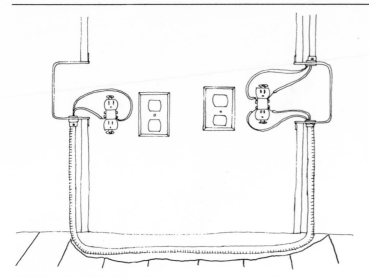

FIGURE 7-28   Running a cable through the floor and across a basement

or sides of joists. Although every old-work job entails special planning, the foregoing procedures illustrate the basic methods of installation.

In difficult situations where wires cannot be fished, or in locations such as garages where appearance is not an essential consideration, surface wiring is utilized, as shown in Figure 7-29. Dual-purpose cable with surface-mounted outlets and switches are easy to install, and the cable is comparatively inconspicuous if it is approximately the same color as the wall. Cable can be painted any desired color. It can be secured to the wall by means of staples, as in concealed wiring. When surface wiring needs protection, it is enclosed in metal raceways, which are available in various sizes and shapes. Figure 7-30 shows the appearance of a metal raceway. Fittings similar to those used in conduit installations are employed in raceway runs; the raceway is never bent or curved, for example. A raceway is strapped to surfaces in the same manner as conduit. Like the cable, the raceway can be painted to match the color of the wall, if desired.

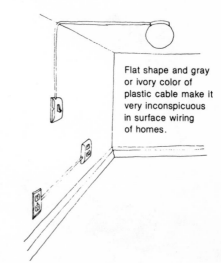

Flat shape and gray or ivory color of plastic cable make it very inconspicuous in surface wiring of homes.

FIGURE 7-29   Example of surface wiring with dual-purpose cable

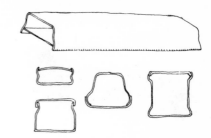

FIGURE 7-30   Metal raceway

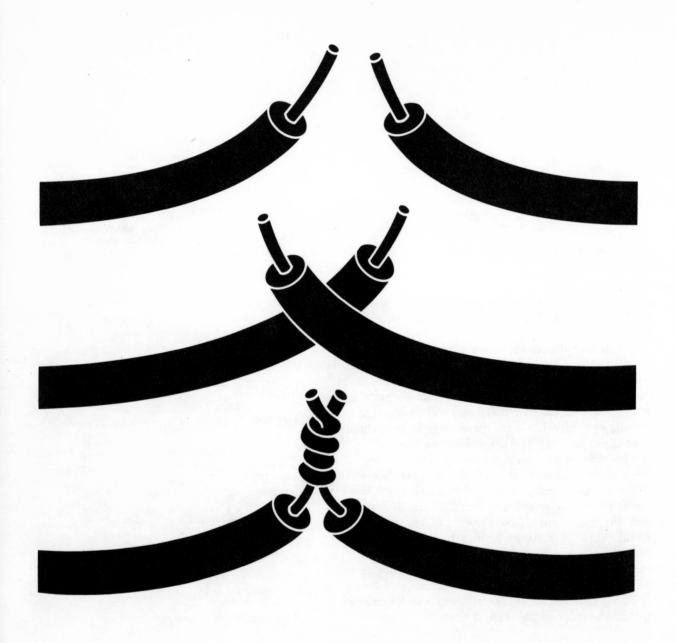

HOW TO INSTALL INDIVIDUAL CIRCUITS 8

GENERAL CONSIDERATIONS

When individual circuits are to be installed, it is important to observe grounding requirements. In other words, cable with a grounding wire must be installed in most (but not in all) new work. Formerly, cable with a grounding wire was run only to receptacle boxes, but code requirements have become more strict with regard to metal boxes. In some situations, cable with a grounding wire must be run to every box; it is advisable to consult with your local public utility or inspection authority in this matter. The bare grounding wire makes connection first at the neutral strip in the service equipment cabinet, as shown in Figure 8-1. When cables run into or out of a fixture outlet, the grounding wires are connected as shown in Figure 8-2. Note that the bare wires all connect to each other and to the metal box. In the case of a receptacle outlet, the bare wires all connect to each other and to the green terminal of the receptacle, as Figure 8-3 illustrates. Solderless connectors are screwed on the ends of wires (See Figure 8-4). In the case of insulated wires, just enough of the insulation is removed to permit attachment of the connector.

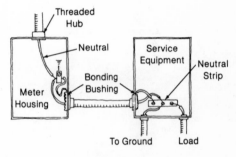

FIGURE 8-1   The grounding wire starts at the neutral strip in the service equipment cabinet

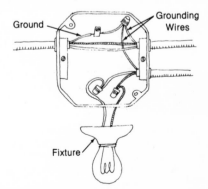

FIGURE 8-2   Grounding connections in a fixture outlet

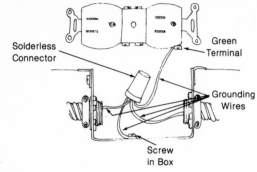

FIGURE 8-3   Grounding connections in a receptacle outlet

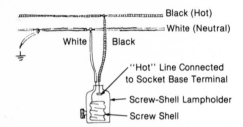

FIGURE 8-5   Color coding of fixture wires

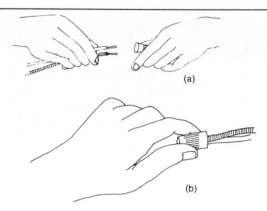

FIGURE 8-4   Installation of solderless connector   (a) Preparation of wires   (b) Cap is screwed on tightly.

Color coding of the black (hot) and white (neutral) wires must be observed when installing a fixture. This ensures that the neutral wire is connected to the screw shell (Figure 8-5) as required by the NEC to reduce the shock hazard in changing bulbs. Note that although the neutral wire is grounded at the service entrance, neutral is never used as a grounding wire. Both the black and the white wires are regarded as power conductors. A grounding wire never conducts electricity unless there is a fault in the wiring system. As noted previously, a cable may consist of a black wire, a red wire, and a white wire, all of which are regarded as power conductors. There are 120 volts between the black wire and the white wire, 120 volts between the red wire and the white wire, and 240 volts between the black wire and the red wire. A three-wire cable may also include a bare grounding wire. Note in passing that a toggle switch is occasionally provided with a green terminal, the same as for a receptacle. It is good practice to connect the grounding wire to the green terminal of a switch, in order to minimize the possibility of shock.

SWITCH CIRCUITING

A ceiling light may be controlled either by a pull-chain switch or by a wall switch. If a wall switch is being installed to control a ceiling light at the end of the cable run, make connections as shown in Figure

8-6(a). Or, if two ceiling lights are to be installed on the same line, one controlled by a pull chain and the other controlled by a wall switch, make connections with three-wire cable as shown in Figure 8-6(b). Note

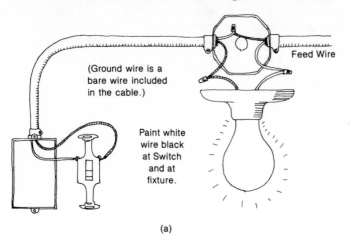

(Ground wire is a bare wire included in the cable.)

Feed Wire

Paint white wire black at Switch and at fixture.

(a)

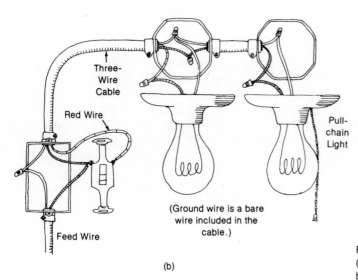

Three-Wire Cable

Red Wire

Pull-chain Light

(Ground wire is a bare wire included in the cable.)

Feed Wire

(b)

FIGURE 8-6   Installation of a wall switch to control ceiling light (a) Switch installed at end of cable run   (b) Switch installed before ceiling outlet

carefully that when a switch is installed at the end of a cable run, the ends of the white wire in the cable must be painted black at the switch and also where the white wire joins the black wire in the fixture box. This black paint is required to maintain consistency of the color code, because cable with two black wires is unavailable. Although the ground wire is indicated separately in Figure 8-6 for purposes of clarity, it is understood that the ground wire runs in the same jacket as the black and white wires. A ground wire is always continuous throughout a wiring system, and a white wire is always continuous. All switching connections are made with black wires or red wires, which in turn are not continuous.

Next, a switch and receptacle outlet can be installed at the end of a cable run as shown in Figure 8-7(a), with the wall switch wired to control the ceiling light. The receptacle is "hot" at all times. If desired, a two-gang box can be used for the switch and receptacle, as shown in Figure 8-7(b). Two receptacles

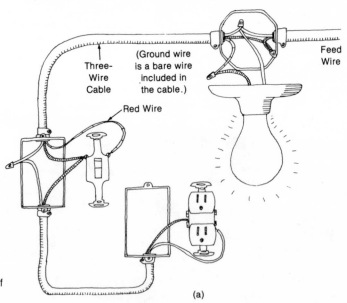

FIGURE 8-7   Installation of wall switch and receptacle at end of cable run   (a) Switch and receptacle in separate boxes

are installed in separate boxes with connections as illustrated in Figure 8-8. For purposes of clarity, the ground wire is not shown; however, in most installations a ground wire will be required, as explained previously. If the receptacles have light and dark terminal screws, connect the white wire to the light terminal, and connect the black wire to the dark terminal. When a sufficiently long section of cable is not available for a run, two sections of cable must be

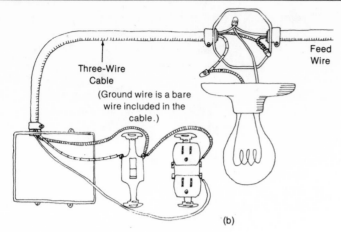

Three-Wire
Cable

(Ground wire is a bare
wire included in the
cable.)

Feed
Wire

(b)

FIGURE 8-7   (b) Switch and receptacle in a ganged box

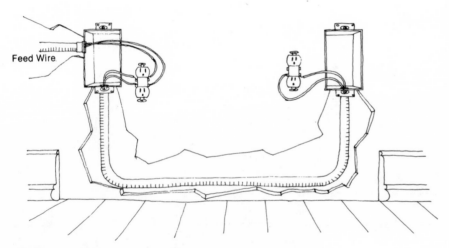

Feed Wire

FIGURE 8-8   Installation of a pair of duplex outlets in separate
boxes

connected together as depicted in Figure 8-9. All connections must be made in junction boxes. The junction
box is covered with a plain plate. If the junction box is
installed in a location where it could be touched, it
should be grounded in the same manner as a switch
box or outlet box.

Next, suppose that two ceiling outlets are to be
independently controlled by two toggle switches in a
two-gang box; the wiring connections are made as
shown in Figure 8-10. Note that both ends of the white

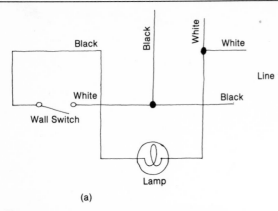

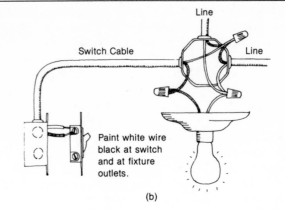

FIGURE 8-9 Ceiling light controlled by wall switch with line continuing past the ceiling outlet box (a) Schematic diagram (b) Pictorial diagram

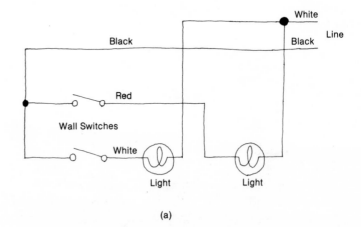

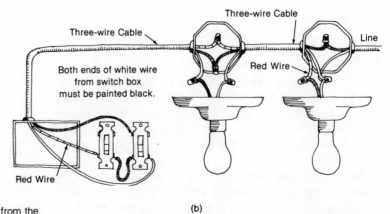

FIGURE 8-10 Two ceiling lights wall-switch controlled from the same switch box (a) Schematic diagram (b) Pictorial diagram

wire from the switch box must be painted black. As noted previously, three-way switches are generally used to control one or more fixtures from different locations, so that the light(s) can be turned on or off by either switch. Figure 8-11 shows how to connect a pair of three-way switches at the end of a cable run to control a light fixture. Color coding must be carefully observed; a three-way switch has a pair of light-colored terminals A and B, to which the red and white wires are connected. The switch also has a dark-colored terminal C to which the black wire is connected. Light-colored terminals often have a bright brass finish, while dark-colored terminals have an oxidized finish. For purposes of clarity, the ground wire has been omitted in Figure 8-11; however, a conventional ground wire must be included in most installations.

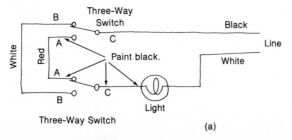

(a)

In case a ceiling fixture is to be controlled with a pair of three-way switches with a receptacle outlet at the end of the cable run, make the connections as shown in Figure 8-12. Only the fixture is in the switching curcuit, and the receptacle is always "hot." A and

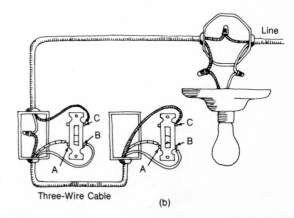

FIGURE 8-11   Light is controlled by two three-way switches, both installed past the lamp   (a) Schematic diagram   (b) Pictorial diagram

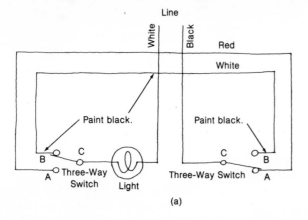

(a)

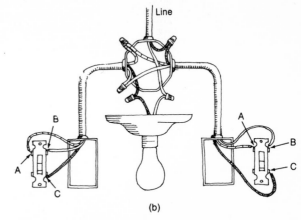

(b)

FIGURE 8-12   Ceiling outlet installed between a pair of three-way switches   (a) Schematic diagram   (b) Pictorial diagram

B denote light-colored terminals, and C denotes a dark-colored terminal. As before, a grounding wire is not included in Figure 8-12, although it will usually be required. When three-way switches are installed, the white wire from the switches must be painted black both at the switch end and at the light outlet end. If a ceiling fixture is located beyond a pair of three-way switches, make connections according to Figure 8-13.

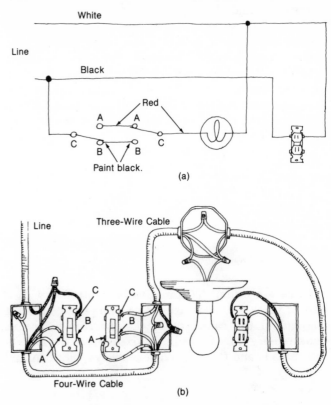

FIGURE 8-13 Ceiling outlet installed beyond a pair of three-way switches with a convenience outlet (a) Schematic diagram (b) Pictorial diagram

Again, if the fixture is to be controlled from three locations, a pair of three-way switches and one four-way switch will be required, with circuit connections as shown in Figure 8-14. As before, A, B, and C denote light- and dark-colored terminals on the three-way switches; AA and BB also denote light-colored terminals not connected to the same wires as A and B.

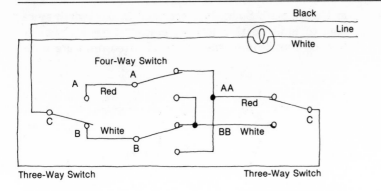

Black
Line
White

Four-Way Switch

A
A
Red
A
AA
Red
C
C
White
BB White
B
B
B

Three-Way Switch

Three-Way Switch

(a)

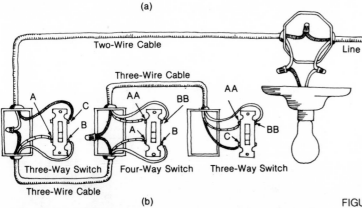

Two-Wire Cable

Line

Three-Wire Cable

A
C
AA
BB
AA
B
A
B
BB
C

Three-Way Switch

Four-Way Switch

Three-Way Switch

Three-Wire Cable

(b)

FIGURE 8-14   Ceiling outlet controlled from three locations
(a) Schematic diagram   (b) Pictorial diagram

Figure 8-15 shows how a pair of three-way switches operate, for all of the switching combinations. Many combinations of three-way and four-way switches are possible; the more usual arrangements are illustrated

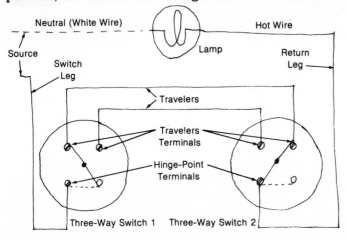

Neutral (White Wire)
Hot Wire

Source

Lamp
Return
Leg

Switch
Leg

Travelers

Travelers
Terminals

Hinge-Point
Terminals

Three-Way Switch 1   Three-Way Switch 2

FIGURE 8-15   Operation of two three-way switches

in Figure 8-16. Solid lines are used in the switch symbols to indicate the path of current flow for the reference condition of the circuit. Dotted lines are used

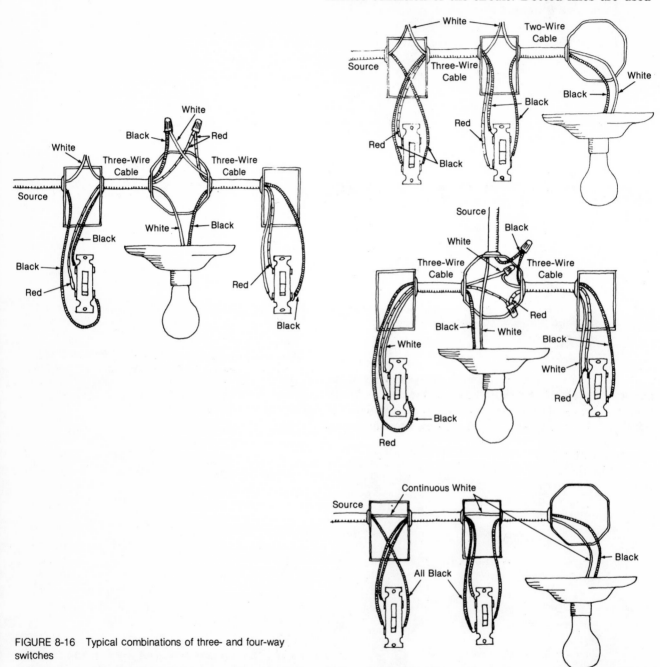

FIGURE 8-16   Typical combinations of three- and four-way switches

to show the paths of current flow for other than the reference condition. In each diagram, the lamps are off, but will be turned on if any switch is thrown to its next position. With the lamps turned on, they can be turned off if any switch is thrown to its next position. There is a limit of two three-way switches in any circuit; however, as many four-way switches can be installed as may be desired.

HEAVY APPLIANCE CIRCUITING

As noted previously, major appliances operate on 240 volts and require individual circuits. Comparatively heavy conductors and large devices are utilized to handle the large circuit currents efficiently. As an example, Figure 8-17 shows a typical power receptacle mounted in a metal outlet box. Figure 8-18 illustrates a larger receptacle and mating plug rated for operation at 50 amperes and 250 volts. An electric range is installed with a three-wire No. 6 cable running to a 50 ampere branch circuit in the main service panel at one end, and to a heavy-duty wall receptacle at the other end (see Figure 8-19). A flexible three-wire cord, or so-called "pigtail" is connected to the range terminals.

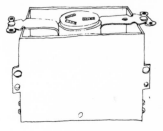

FIGURE 8-17   Power receptacle mounted in a metal outlet box

FIGURE 8-18   Power receptacle and plug rated for operation at 50 amperes and 250 volts

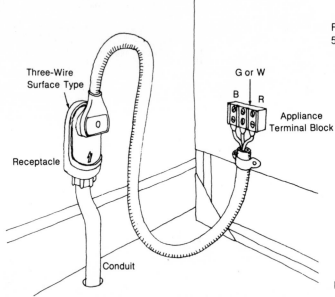

Three-Wire Surface Type

Receptacle

Conduit

G or W

B   R

Appliance Terminal Block

FIGURE 8-19   Typical range or dryer receptacle installation

The black wire connects to B, the red to R, and the white or green wire to G or W. Note that the neutral wire is color-coded white or green in a 240-volt circuit. The frame of the range is grounded by connection to the neutral terminal. In other words, a bare ground wire is not utilized in a 240-volt system. A switch is not included in the range pigtail circuit; to disconnect the range, the plug is pulled out of the receptacle.

A dryer is installed in the same manner as a range, except that it operates from a 30-ampere 240-volt branch circuit. However, a high-speed dryer is operated from a 50-ampere branch circuit. It is advisable to consult with your local inspection authority to determine any special local requirements before installing heavy appliance circuits; wire size and grounding details may differ somewhat from one locality to another. In regard to water heaters, the method of wiring will be specified by your local public utility. A double-element heater is advisable for a large family, because it can provide more hot water at an even temperature. If the service-entrance panel has an unfused tap for connection to a water heater, an indoor fused safety switch must be installed, as shown in Figure 8-20. Note that this is a completely separate circuit

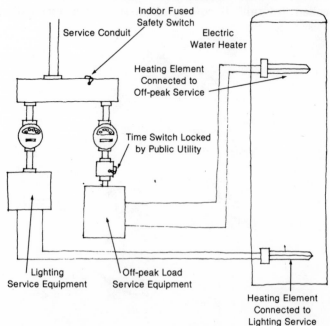

FIGURE 8-20   Two-element water heater connected to off-peak-load service and to lighting service

from the line to the range receptacle. The public utility may install a separate kwh meter and a time switch for a water heater, so that low operating cost can be provided by off-peak load service. Of course, if full capability of the water heater is needed during peak-load hours, a separate meter and time switch will not be installed.

Three widely used types of heavy-duty appliance receptacles are shown in Figure 8-21. These are wall-type receptacles and mate with the same standard heavy-duty plug. Note that if conduit is installed in a heavy-appliance circuit, the neutral wire must be white. It some localities, service-entrance cable is utilized, and the neutral wire is bare. On the other hand, a water heater does not have a neutral wire, because it operates in a 240-volt circuit, whereas a range or a washer-dryer operates in a 120/240-volt circuit. In other words, a water heater has two connecting wires, whereas a range or a washer-dryer has three connecting wires. If the installation of an air conditioner is planned, the local public utility should be consulted to determine whether power for the additional load is available, and to determine the ampacity of the service

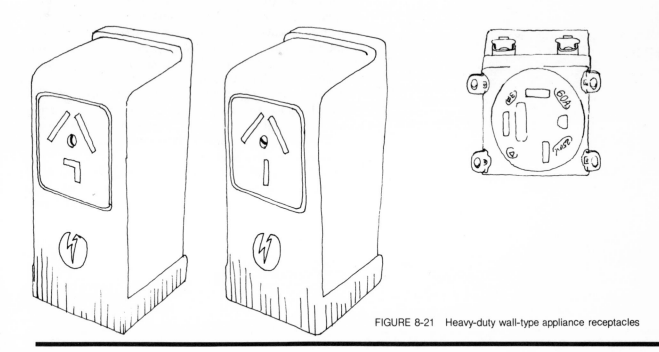

FIGURE 8-21   Heavy-duty wall-type appliance receptacles

conductors that will be required. An air conditioner operates on a three-wire 240-volt circuit with at least a 100-ampere rating.

LOW-VOLTAGE CIRCUITING

Practically all residences have some low-voltage circuiting for operation of doorbells, chimes, or buzzers. A 10- to 16-volt stepdown transformer is utilized as a power source (Figure 8-22). Doorbells or buzzers generally operate on 10 volts, whereas chimes usually operate on 16 volts. No. 18 bell wire is widely used in low-voltage circuit installation. The stepdown transformer is typically installed on an outlet box in the basement or utility room. The primary winding is permanently connected to a 120-volt line. Doorbells and chimes are installed in any desired location. Bell wire is run over exposed surfaces, under mouldings, or behind baseboards, or it may be fished through the walls. Concealed wiring is always installed in new work. Insulated staples are used to attach bell wires to surfaces.

In a doorbell-buzzer circuit, the buzzer may announce a caller at the back door, whereas the bell announces a caller at the front door. Circuits are shown in Figure 8-23. Note that a buzzer and bell are sometimes combined into a single unit. Chimes are usually designed to sound one note for the back door, and two notes for the front door.

Branch circuits for bells, chimes, and buzzers are not fused, because the regulation of a bell-ringing transformer is comparatively poor. Furthermore, if an accidental short-circuit should occur, the current flow is limited and no fire hazard results.

Pushbutton switches can be obtained with a built-in pilot light; this design makes the pushbutton visible in the dark, so that a caller does not need to fumble for the button.

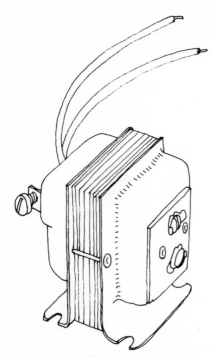

FIGURE 8-22 Stepdown transformer used for doorbells and buzzers

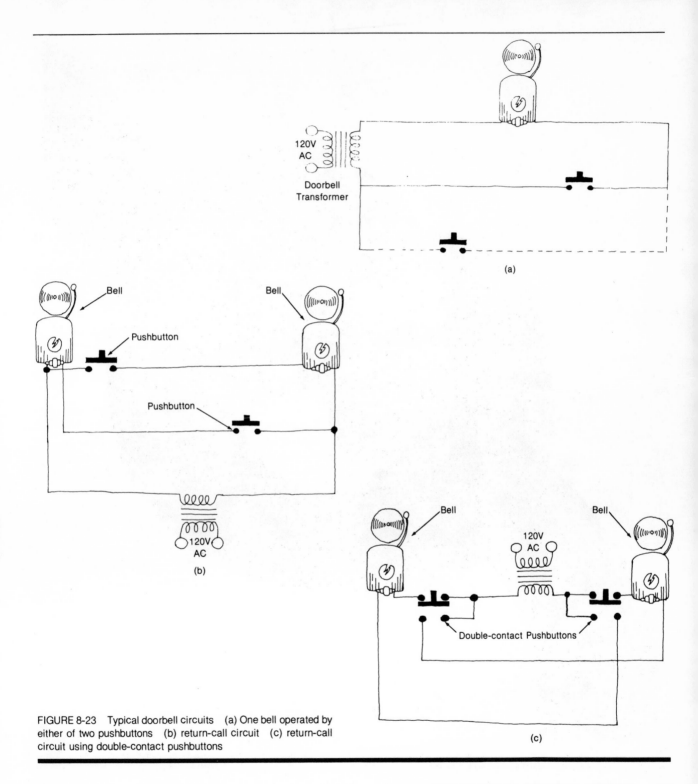

FIGURE 8-23   Typical doorbell circuits   (a) One bell operated by
either of two pushbuttons   (b) return-call circuit   (c) return-call
circuit using double-contact pushbuttons

TROUBLE-SHOOTING WIRING SYSTEMS 9

GENERAL CONSIDERATIONS

Circuit breakers will trip and fuses will blow occasionally in practically all wiring systems. When the cause of overload is corrected the circuit breaker (Figure 9-1) can be reset, and will remain so; or the blown fuse can be replaced and will hold. A blown fuse can be easily located with a neon tester. Referring to Figure 9-2, touch terminal 1 with one of the neon terminals, and touch terminal 3 with the other neon terminal. If the lamp does not glow, the fuse is blown; but if the lamp does glow, the fuse is good. If the test shows the fuse is good, the neon tester is applied to terminals 2 and 3. Proceed, if necessary, to check terminals 3 and 4, and then check terminals 3 and 5. In this example, the lighting and appliance circuits are tested at terminals 6 and 9, 8 and 9, 10 and 9, and 12 and 9. The main fuse is tested at terminals 7 and 9 and at 11 and 9.

In case a fuse blows repeatedly, there is most likely a defective appliance or light in the circuit. There may be a defective lamp cord or appliance cord that draws a short-circuit current. To localize the defect, first turn

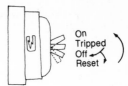

FIGURE 9-1   Operating handle positions for a circuit breaker

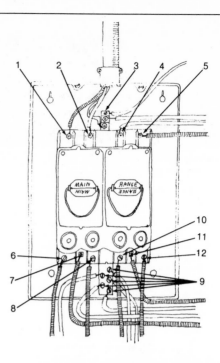

FIGURE 9-2   Test points to check with a neon tester for blown fuses

off all the ceiling lights, turn off all wall switches, and disconnect all cords, lamps, and appliances on the branch circuit. Instead of inserting a new fuse, screw a 100-watt light bulb into the fuse socket. If the bulb is dark, there is no defect in the circuit conductors; on the other hand, if the bulb lights up, there is a short-circuit somewhere along the conductors. If the bulb is dark, plug the appliances and lamps in, one by one, until the bulb lights up. In other words, if the bulb in the fuse socket lights up, but the lamp that you have plugged in remains dark, the trouble will be found in the plug, cord, or socket of the lamp that you have plugged in. Note that on rare occasion, a light bulb will burn out in such a way that it leaves a short-circuit.

Now, if the bulb in the fuse socket lights up when all the switches are turned off and all the cords are disconnected from the branch circuit, it follows that there is a short-circuit at some point in the cable run. In nearly all situations of this sort, you will find the short-circuit in a box. For example, careless stripping of a black wire or red wire may permit contact with a grounding wire, a white wire, or the box. Such defects of installation are usually caught during inspection, but there are exceptions. In case the short-circuit is not in a box, the most likely fault is a nail that has been accidentally driven into the cable at some point along the run. Correction of the trouble involves not only removal of the nail, but also replacement of the defective cable section. Note that if the branch circuit is protected by a breaker instead of a fuse, a light bulb is not used as an indicator. Instead, the operating handle of the breaker is thrown to the "on" position each time the circuit is tested.

INADEQUATE WIRING INSTALLATION

Troubleshooting often involves an inadequate wiring installation, particularly in old work. Symptoms of an inadequate wiring system are lights that dim when appliances are turned on, slow warmup and low heat on appliances such as toasters or waffle irons and heaters, frequent blowing of fuses or tripping of circuit breakers, TV pictures that become narrow or hazy or lose sync when appliances are turned on, and motors that slow down under normal work load. Correction of an inadequate wiring system usually involves only the provision of a few additional branch circuits. An add-on panel such as the one illustrated in Figure 9-3 can

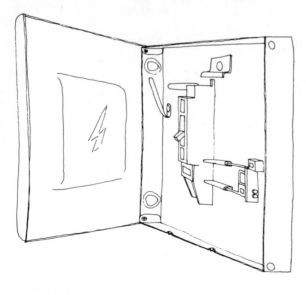

FIGURE 9-3   A 20-amp circuit breaker installed in an add-on branch-circuit-breaker panel

be installed and connected to the power takeoff lugs and the neutral strip in the service-entrance panel. From the add-on panel, a new branch circuit can be run into the room where the existing circuit is being overloaded (usually the kitchen). The new branch circuit typically supplies additional appliance outlets, so that the load on the existing circuit(s) can be reduced.

Sometimes a wiring system appears to be inadequate because of a high-resistance connection. That is, if there is a high-resistance contact to a receptacle, a lamp plugged into that receptacle will be dim. The clue to the poor contact is that if the lamp is plugged into another receptacle on the same branch circuit, the lamp lights up normally. A high-resistance contact is a fire hazard, because the power loss in the contact generates excessive heat. Although the insulation of the cable inside the box may catch fire, the box enclosure provides good protection against setting the wall on fire. A poor contact sometimes becomes intermittent, and a lamp may go on and off erratically when the plug is twisted in the receptacle. In this situation, the poor contact may be found either in the plug or in the receptacle.

LOST NEUTRAL SYMPTOMS

Occasionally, a wiring installation is plagued with the symptoms of a lost neutral. For example, one lamp may glow too brightly while other lights glow dimly. As various lights are switched on and off, the other lamps on two branch circuits vary in brightness. Figure 9-4 shows the basis of these symptoms. Note that the current through all four lamps flows into or out of the neutral wire. If all the lamps are the same, and all four lamps are turned on, there is no current flow through the neutral wire into the center-tap of the secondary winding. On the other hand, suppose two lamps are turned on in one branch circuit, and one lamp is turned on in the other branch circuit. Then there is an unbalanced current condition, and the current of one lamp flows through the neutral wire into the center-tap of the secondary winding. Now, suppose that the neutral wire becomes broken at some point between the lamps and the ground connection. The result will be symptoms of a lost neutral.

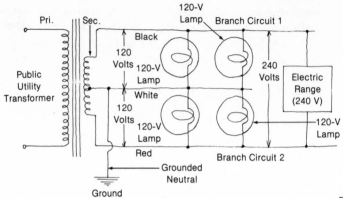

FIGURE 9-4   Plan of a three-wire system for a residential installation

With an open-circuited neutral wire like the one in Figure 9-4, the electric range will operate normally; yet, the lamps will operate normally only in case both branch circuits are loaded equally. For example, if two lamps are turned on in branch circuit No. 1, and both lamps are turned off in branch circuit No. 2, none of the lamps glow. Then if one lamp is turned on in branch circuit No. 2, it will light up brilliantly and soon burn out; the lamps in branch circuit No. 1 will glow dimly until the first lamp burns out, and then they, too, will be dark. When an open-circuited neutral wire is being checked out, turn off the power at the service panel and inspect the connection of the neutral wire at the neutral strip. Also inspect the neutral connection in the add-on panel, if one has been installed. Although poor connections in panels are comparatively rare, this is an easy fault to find if it happens to occur. If you must search further, however, check the neutral connections in the various boxes of the branch circuits that are affected. Another possibility of trouble is severance of the neutral wire along a run by a nail that has been accidentally driven through the cable.

INSTALLATION ERRORS IN WIRING

After a wiring system has been installed, it should be checked out before power is applied from the service. Most electricians use a doorbell and a dry battery to check continuity. It is extremely important to make certain that a wiring installation is "dead" before making continuity tests. After the wiring has been installed, temporarily twist together the wires that will be spliced. Where a switch is to be installed, connect the wires together to simulate the "on" condition of the circuit. Also connect the black, white, and red wires of the branch circuit together at the service end, as shown in Figure 9-5. Then, if there are no errors in the wiring, the bell will ring when the test leads are applied to each pair of wires in the outlet boxes. After checking out the wiring, remove the temporary connections and proceed to finish the installation. In a finished installation, you can check out the continuity of the ground system between the neutral wire and each box.

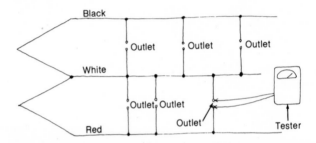

FIGURE 9-5   Wiring installation arranged for continuity tests

IDENTIFYING BRANCH CIRCUITS

When making an analysis of an existing wiring installation, one of the important considerations is the identification of each branch circuit in a system. This is easily accomplished in a normally operating installation. Open all of the branch circuit breakers (or fuses) except one. Then, check each fixture and receptacle to see if it is "live" or "dead." All the live outlets are wired into the branch circuit that was not opened. Then you can identify each of the other branch circuits in turn by repeating the procedure. It is essential to make certain that all the switches in the branch circuit under test have been turned on. Checking of receptacle outlets is facilitated by use of a light bulb with a plug-in type of socket.

CORRECTING WIRING ERRORS

When a wiring error has been made, you must evaluate test results in order to determine the nature of the error and its location. This determination requires detailed consideration of the circuitry and a logical processing of the test results. Continuity tests generally provide the necessary data. When open circuits are observed where short circuits are expected, or short circuits are observed where open circuits are expected, you must consider these abnormal conditions with respect to the circuitry that is involved. For example, if a circuit is overloaded and the fuse has been defeated with a penny or other cheater, an open circuit is likely to result from a melted conductor along the run. Of course, this situation also causes disastrous fires upon occasion. A diagram of the complete wiring installation may be available; if not, take time out to draw up a complete diagram. Then, mark outlets that have abnormal short circuits with red ink, and mark those with abnormal open circuits with green ink. A study of the mark-up will then lead to recognition of the error(s). Note that abnormal open circuits do not always result from wiring errors.

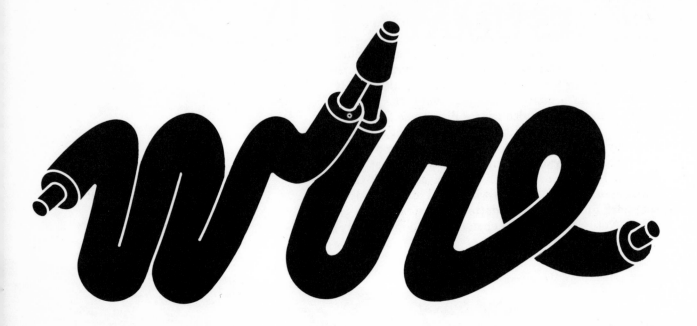

ELECTRICAL COST ESTIMATING 10

ELECTRICAL ESTIMATING

In conclusion it must be noted that an electrical wiring project involves both material and labor costs. The cost of wire, boxes, switches, outlet receptacles, fixtures, GCFIs, and service entrance equipment can be calculated after planning the project. The labor cost depends upon both the prevailing wage scale and the efficiency of the electrician. For example, a highly skilled electrician might rough-in a typical tract house in one day. Then, after the walls and ceilings have been finished, the electrician might complete the installation of switches, receptacles, wall plates, and fixtures in one more day. On the other hand, a moderately skilled electrician might take twice as long to do the same work. An electrical contractor knows how much time will be required for an electrician in his employ to complete a typical project.

Instead of making a detailed breakdown on residential wiring projects the usual procedure is to calculate the cost in terms of the number of boxes that are to be installed, plus certain supplementary costs. As an illustration, an electrical contractor will allow a little over one hour for installation of each box. A supplementary charge equivalent to 11 or 12 hours is typically made for service installation. However, if an underground service installation is to be made, this supplementary charge is increased accordingly. Supplementary charges are also made for each 240-volt outlet, because material costs are somewhat greater than for 120-volt outlets. In the case of a dryer or electric range, a separate charge is made for cable if the run is over ten feet, because this type of cable is comparatively heavy and costly.

In estimating a wiring project, a contractor might charge an amount that allows about $18 per hour for labor. In turn, the contractor realizes about $7 per hour profit on the electrician's labor. The contractor also takes a profit on material, inasmuch as he purchases fixtures, for example, at wholesale cost and resells them at the retail price. Many electrical contractors operate retail lighting and appliance stores in combination with their contracting business. Note that wages cited are only hypothetical examples and wage scales in metropolitan areas are usually higher than in rural areas.

EXERCISE 1

Electrical Cables and Wires

This exercise provides familiarity with typical electrical cable and wires. Obtain samples of Type NM 14-2 nonmetallic sheathed cable, 14-2 G nonmetallic sheathed cable, 12 Type TW wire, and 2/7 Type TW Oil and Moisture Resistant stranded cable.

1. Remove the insulation from the end of each wire and cable with a pocket knife, being careful not to nick the conductor.
2. Observe that the Type NM 14-2 cable contains two No. 14 insulated wires, whereas the Type 14-2 G cable contins a grounding wire in addition to the pair of No. 14 insulated wires.
3. Note that the 2/7 TW stranded cable is equivalent to a solid No. 2 conductor, but that it is made up from 7 smaller conductors.
4. Using a standard wire gage, measure the size of each conductor in the sample cables and wires. Check to determine whether the grounding wire in the 14-2 G cable is the same size as the insulated conductors.
5. Observe the color coding of the insulated wires in the cables.

EXERCISE 2

Cable Requirements

This exercise provides practice in calculation of cable requirements. With reference to the wiring layouts in Figure 4-4(a), estimate approximately:

1. The number of feet of cable required for each circuit.
2. The total number of feet of cable required for the 120-volt circuits.
3. The total number of feet of cable required for the 240-volt circuits.
4. The total number of feet of cable required for the complete wiring system.

Also determine or estimate the following requirements:

1. The number of 120-volt duplex outlets required.
2. The number of toggle switches that will be necessary, assuming that five of the lights will be provided with three-way switches.
3. The number of 240-volt outlets required.
4. The total number of outlet and switch boxes required for the complete wiring system.

EXERCISE 3

Feeder Ampacity

This exercise provides practice in calculating required feeder ampacity. Make the indicated calculations, and answer the final question.

1. A single-family residence has a floor area of 1500 square feet, exclusive of an unoccupied cellar, an unfinished attic, and open porches. A 12-kw range is installed.

2. The general lighting load is calculated at 3 watts per square foot.

$$1500 \times 3 = \underline{\hspace{2cm}} \text{ watts}$$

3. The minimum number of branch circuits required, according to NEC regulations, is as follows:

   General lighting load, 4500/115 = 39.1 amp; three 15-amp 2-wire circuits will be required, or two 20-amp 2-wire circuits.

   Small appliance load will require two 2-wire 20-amp circuits.

   Laundry load will require one 2-wire 20-amp circuit.

4. The required feeder ampacity is determined as follows:

   Calculated load
   General lighting .... 4500 watts
   Small appliance load .... 3000 watts
   Laundry .... 1500 watts
        Total (without range) .... 9000 watts

   Of this calculated load, the NEC stipulates that the first 3000 watts shall be calculated at 100 percent, and the remainder at 35 percent. Or, the net calculated power demand (without the range load) is:

   $$3000 + 2100 = \underline{\hspace{2cm}} \text{ watts}$$

   Next, the NEC stipulates that the 12-kw range load shall be calculated at 8000 watts.

   In turn, the net calculated feeder power demand becomes:

   $$5100 + 8000 = \underline{\hspace{2cm}} \text{ watts}$$

   This power demand corresponds to an ampacity of:

   $$13100/230 = \underline{\hspace{2cm}} \text{ amperes}$$

   Will a 100-ampere service be adequate, or must a 200-ampere service be installed?

EXERCISE 4

Making Pigtail Splices

This exercise provides practical experience in making pigtail splices and in forming an Underwriter's knot. Obtain a damaged section of two-conductor cable (an undamaged section could also be used), a pair of diagonal cutters or scissors, pliers, and plastic tape. Also obtain a section of lamp cord and a standard plug for a receptacle outlet.

1. As shown in the diagram below, remove the outer insulation for about 2 inches from the damaged point in the cable.
2. Cut the conductors at the damaged point, and strip about an inch of insulation from the end of each conductor.
3. Make pigtail splices, and tighten with a twist of the pliers, if necessary.
4. Wrap each exposed pigtail splice with plastic tape.

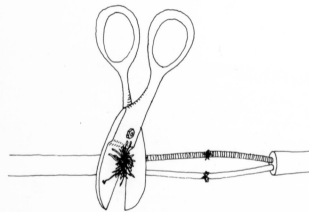

1   Cut away damaged wire.

2   Make pigtail splice in each wire.

3   Wrap each wire with tape.

Next, practice making an Underwriter's knot, and connecting the ends of a lamp cord to the terminals of a receptacle outlet plug.

1. Remove the outer braid from the end of the cord, for a distance of approximately 1½ inches.
2. Insert cord through plug.
3. Tie a knot in the pair of conductors as shown in the diagram below, and tighten the knot at the end of the braid.
4. Strip the insulation from the ends of the conductors for about 1/2 inch back.
5. Pull the cord so that the knot goes down into the recess between the prongs of the plug.
6. Connect the bare ends of the conductors under the terminal screws of the prongs.

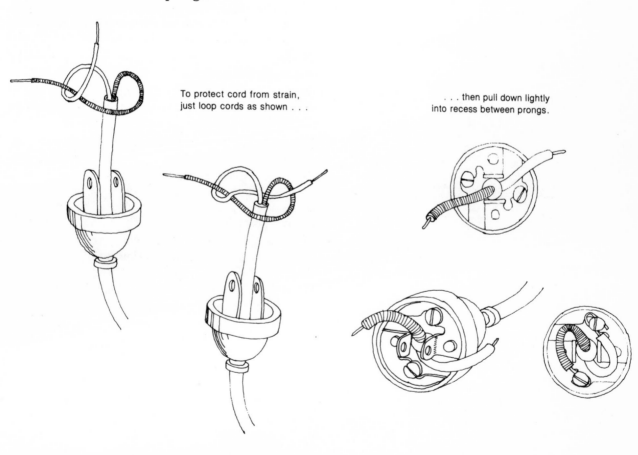

To protect cord from strain, just loop cords as shown . . .

. . . then pull down lightly into recess between prongs.

EXERCISE 5

Converting 2-Wire to 3-Wire Receptacle

This exercise provides a practical experience in converting from 2-wire to 3-wire receptacle operation in old work.

Obtain a metal outlet box, a duplex receptacle outlet of the two-prong type, an outlet adapter, and a three-prong grounding plug.

1. Observe that the three-prong grounding plug will not fit into a two-prong receptacle outlet.
2. As shown in the diagram below, a three-prong grounding plug will fit into an outlet adapter, which in turn will fit into a two-prong receptacle outlet. Check out this arrangement.
3. Observe that the outlet adapter has a grounding wire with a lug.
4. Connect the lug of the grounding wire under one of the screws that is used to mount the receptacle on the box, and tighten the mounting screws.
5. Note that this arrangement provides a ground connection to the grounding wire of the adapter, provided that the box is grounded.
6. In case a box is not grounded, it is necessary to connect the grounding wire of the adapter to a ground point, such as a water pipe.
7. Could a ground wire be "fished" to a water-pipe location in a frame-construction building (in most cases)?
8. What type of building construction would require that a ground wire to be run to a water pipe in a surface metal raceway?

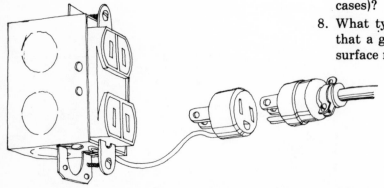

EXERCISE 6

Ganging Metal Boxes

This exercise provides practical experience in ganging metal boxes. Obtain a pair of metal switch boxes.

1. Observe the construction of the metal boxes.
2. Remove the wall end of each box, as shown in the diagram below.

3. Fit the open sides of the boxes together.

4. Tighten the screws, and secure the boxes together as a single unit.
5. Does the ganged box require a pair of single wall-plates, or a double-sized wallplate?
6. Could you install a toggle switch and a duplex receptacle outlet in the ganged box?
7. Could you install a pair of duplex receptacle out-lets in the ganged box?
8. Is it possible to gang three or four metal boxes?
9. List several combinations of devices that could be installed in a three-box gang.

Making Simple Connections

This exercise provides practical experience in making simple connections. Obtain a standard junction box with cable clamps, a standard switch box with cable clamps, an SPDT switch, five short sections of 14-2 NM cable, and three wire nuts.

1. Strip the insulation from the conductors at one end of each cable, leaving 3/4 inch of conductor exposed.
2. Insert three of the cables into the junction box as shown below, and tighten the cable clamps.

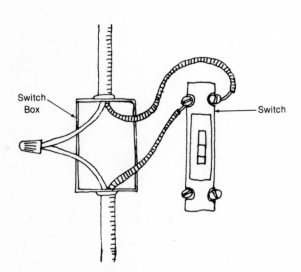

Junction Box

Wire Nut

Switch Box

Switch

3. Connect (twist) the black wires together, and connect the white wires together.
4. Place a wire nut over each of the connections, and turn it up tightly.
5. Insert the remaining two cables into the switch box as shown below, and tighten the cable clamps.

6. Connect the white wires together, using a wire nut.
7. Connect the black wires to the switch terminals. Make sure that the loops of the black wires are made in the same direction that the terminal screws tighten.

EXERCISE 8

Ground Clamps and Straps

This exercise provides familiarity and practical experience with a conduit ground clamp terminal assembly.

Obtain a conduit ground clamp of the type illustrated and a ground wire. Also obtain a short section of ordinary water pipe.

1. Inspect the conduit ground clamp, and observe the provisions for securing both the pipe clamp and ground wire clamp of the assembly. Remove the large screw to separate the assembly.
2. Scrape or sandpaper the surface of the water pipe where you will mount the clamp.
3. Slide the pipe clamp over the water pipe, and tighten it securely.
4. Insert the ground wire into its portion of the assembly and tighten securely.
5. Now reassemble the terminal using the large screw removed previously.
6. Approximately what size water pipe does the ground clamp accommodate?
7. What size conduit does the ground clamp accommodate?

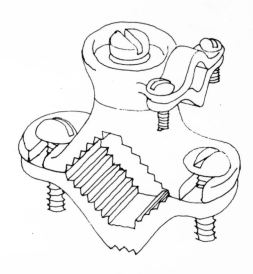

EXERCISE 9

Doorbell Transformer

This exercise demonstrates the characteristics of a doorbell (or chime) ringing transformer.

Obtain a doorbell transformer, an AC voltmeter, an AC ammeter, and a rheostat of the wirewound type with approximately 100 ohms maximum resistance.

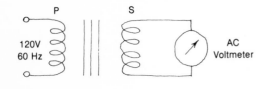

1. Connect the A.C. voltmeter as shown.
2. With the rheostat disconnected from the secondary terminals, measure the open-circuit secondary voltage with the AC voltmeter.

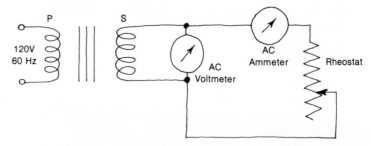

3. Connect the rheostat and AC ammeter into the secondary circuit, and observe the secondary voltage and current for five different settings of the rheostat, as follows: Maximum resistance, 3/4 of maximum resistance, 1/2 of maximum resistance, 1/4 of maximum resistance, and at minimum resistance settings.
4. The change of secondary voltage under increasing secondary current demand is called the regulation of the transformer. If the regulation were 100%, there would be no change in secondary voltage at maximum current demand.
5. Is the regulation of the transformer 100 percent?
6. Let the transformer operate for 20 minutes with short-circuited secondary terminals. Does the transformer heat up excessively?
7. What is the advantage of poor regulation in a doorbell transformer?

EXERCISE 10

Planning Lighting Installations

This exercise provides practice in planning lighting installations. The layout below shows a simple plan for a living-room wiring system. Using this as a starting point, redraw the layout, and elaborate it to obtain a good balance between direct and indirect sources of illumination. Consider the possibilities of employing valance lighting near the ceiling, cornice lighting, cove lighting, or recess lighting. Note that four table lamps have been indicated for supplementary lighting in the original layout. If you feel that one or more pole lamps would be helpful, indicate their locations in your layout drawing.

EXERCISE 11

Installing a Three-Way Light Circuit

This exercise provides practical experience with three-way circuiting. A mock-up or "breadboard" arrangement is utilized.

Obtain two switch boxes, two three-way switches, a fixture outlet box, several feet of three-conductor nonmetallic sheathed cable, a length of two-conductor nonmetallic sheathed cable, a small lighting fixture, five wire nuts, and a receptacle plug.

1. Arrange the hardware items on a table top as depicted in the diagram below.
2. Cut the cables to suitable lengths, and strip the insulation from the ends of the cables.
3. Make the connections as shown in the diagram.
4. Connect the black, white, and red wires as shown. Terminals A and B are the light-colored terminals to which red and white wires must be connected. Terminal C is dark colored, to which the black wire must be connected.
5. Paint the ends of the white wire from the switches black, both at the switches and at the light outlet.
6. Plug the feed wire into a 117-volt output, and check the operation of the three-way switches. If the circuit does not operate correctly, check your connections carefully to locate the wiring error.

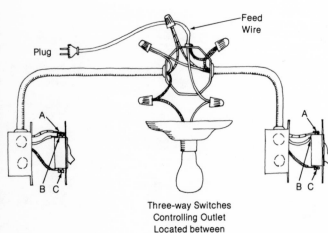

Three-way Switches
Controlling Outlet
Located between
the Switches

EXERCISE 12

Mounting Work Boxes

This exercise provides a practical experience in mounting old-work boxes, and the use of metal box supports.

Obtain a Grip-Tite box, a pair of metal box supports, a piece of wallboard, an auger, and a keyhole saw.

1. Mark the rectangular outline of the box opening on the wallboard with pencil.
2. Use the auger and keyhole saw to make a cutout in the wallboard that will pass the box.
3. As shown in Figure 5-20, push the box into the cutout, with the front brackets resting against the wallboard.
4. Insert a metal support on each side of the box. Work the supports up and down until they fit firmly against the inside surface of the wall. Then bend the projecting ears as shown to fit around the box.
5. Tighten the side screws on the Grip-Tite box to bring the side brackets up snug against the wall.

Losing the Neutral

This exercise provides familarity with troubleshooting the wiring defect called "losing the neutral."

With reference to the diagram below, consider the circuit change that occurs when the neutral wire is broken at point X, and answer the following questions:

1. Do lamps L_1, L_2, L_7, and L_8 operate normally when their switches are closed?
2. If S_3 only is closed, does L_3 light up?
3. If both S_3 and S_9 are closed, do L_3 and L_9 light up?
4. In case S_3, S_9, S_{10}, and S_{11} are closed, do L_3, L_9, L_{10}, and L_{11} light up?
5. When L_3, L_9, L_{10}, and L_{11} are turned on, do they all glow with the same brightness?
6. Discuss the relative brightness of the lamps when all the switches are turned on, if the upper row of lamps are 100-watt types, and the lower row of lamps are 40-watt types.
7. Explain how the location of a break in a natural wire could be approximately located by means of switching tests.

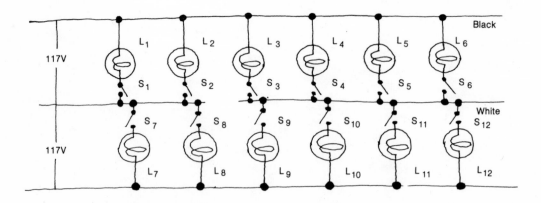

EXERCISE 14

Installation of a Switch Box

This exercise provides practical experience in the installation of a switch box. Referring to the illustration below, proceed as follows:

1. Nail the switch box to the stud, with the front edge protruding sufficiently to be flush with the front surface of the panelboard when the wall is finished.
2. Insert cables through knockouts in the box, and secure the cables with the cable clamps.
3. Cutout is made in the panelboard to pass the box, and panelboard is nailed to the studding.
4. Connect wires to the switch.
5. Secure the switch in place by means of the two machine screws that are provided.
6. Secure the switch wallplate to switch by means of the pair of machine screws that are provided.
7. Repeat the foregoing six steps, using a switch box with a stud mounting bracket. This type of box is illustrated below.

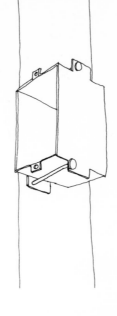
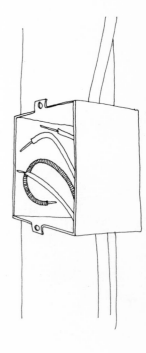

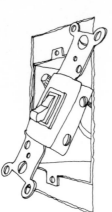
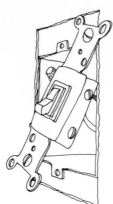
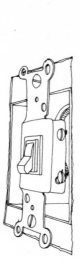

Identifying Branch Circuits

This exercise provides a practical approach to the problem of identifying the branch circuits in an existing wiring system.

The diagram below depicts a circuit-breaker cabinet with a main breaker at the top, and 12 single-pole branch circuit breakers below. From what you have learned about branch circuits, consider each of the following questions, and give a logical answer:

1. If the main breaker is turned off, how many of the branch circuits remain "alive"?
2. If the main breaker is turned on, and 11 of the branch circuit breakers are turned off, how many of the branch circuits remain "alive"?
3. How many branch circuits are there altogether in this example?
4. Is it possible to switch any one of the branch circuits off and on with more than one of the branch circuit breakers?
5. If you were analyzing an old-work wiring installation, suggest a procedure whereby you could identify the fixtures, outlets, and switches that are connected into each of the branch circuits.

EXERCISE 16

Branch Circuit Requirements

This exercise provides practice in evaluating branch-circuit requirements. With reference to Figure 4-5, make the following calculations:

1. Five general-purpose 15- or 20-ampere circuits are specified in the diagram. If all of the lamps and appliances shown on these circuits were turned on at the same time, what would be the total power consumption? What would be the total current demand? What is the minimum number of 15-ampere circuits that could accommodate this demand?

2. Two 20-ampere kitchen-appliance circuits are specified in the diagram. If all of the appliances shown on these circuits were turned on at the same time, what would be the total power consumption? What would be the total current demand? Could two 20-ampere circuits accommodate this demand? Is it at all likely that all of the appliances would be operated at the same time?

3. Calculate the maximum power consumption and the maximum current demand for branch circuits 8 and 9. Could the ampacity of either circuit be exceeded if all the indicated loads were turned on at the same time?

4. Calculate the maximum current demand on branch circuits 10 and 12.

5. Calculate the maximum current demand on branch circuit 11.

6. Calculate the maximum current demand for branch circuits 13 and 15.

7. Calculate the current demand for branch circuits 14 and 16.

8. Calculate the approximate current demand for branch circuits 17 and 19.

9. Calculate the current demand for branch circuit 22.

EXERCISE 17

Conduit Bending

This exercise provides familiarity and practical experience with conduit. Obtain three sections of conduit approximately 5 feet in length, and a conduit bender.

1. Use the conduit bender to make a right-angle bend in a section of conduit. Keep your foot on the conduit bender while applying leverage to the handle. Try to make the radius of the bend approximately eight times the inner diameter of the conduit. (This is the minimum radius permitted by the NEC).

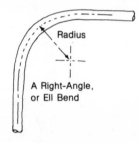

A Right-Angle, or Ell Bend

2. Make an offset bend in another section of conduit. Try to form a throw of 6 inches, and a length of offset of 10 inches.

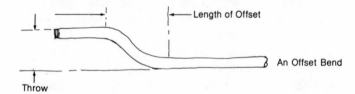

Length of Offset

An Offset Bend

Throw

3. Make a saddle bend in another section of conduit. Try to form a saddle 2 feet in length, with 6-inch offsets.

Offset

Offset

A Saddle Bend

EXERCISE 18

Lighting-Circuit Project

This exercise provides practice in planning branch circuits for residential wiring systems. With reference to the floor plans shown below, calculate how many 15-ampere general-purpose circuits will be required, how many 20-ampere kitchen-appliance circuits will be necessary, and how many 240-volt circuits will be required. Assume that 100-watt lamps are to be used throughout.

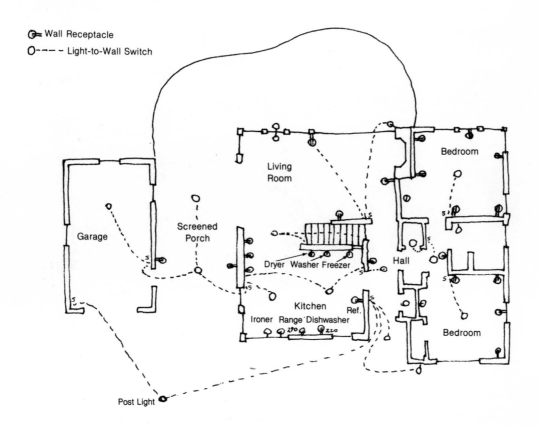

⊖= Wall Receptacle

O---- Light-to-Wall Switch

EXERCISE 19

Estimation of Project

With reference to the following electrical wiring layout, calculate the approximate cost of the project, using the cost figures exemplified in the text. For the purpose of rough estimation, compute the cost of each 240-volt outlet at 50 percent more than for a 120-volt outlet.

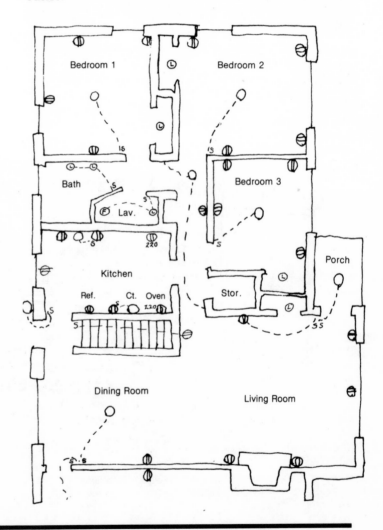

GLOSSARY

A.C. Alternating current

aerial. Wires supported above ground and used for receiving or sending electrical waves

advance wire. An alloy of copper and nickel used for electric heating units

alternating current. An electric current that reverses its direction of flow at regular intervals

alternation. One vibration instead of a cycle; one-half a cycle of alternating current

American wire gauge. Gauge used for designating the sizes of solid copper wires used in the United States. Formerly called Browne and Sharpe; best gauge

ammeter. The instrument that indicates the rate of flow of electricity through a circuit

ampere. The practical unit that indicates the rate of flow of electricity through a circuit

ampere-hour. The quantity of electricity delivered by a current of one ampere flowing for one hour; used in rating storage batteries

ampere-hour meter. An instrument that registers or records the number of ampere-hours of electrical energy that have passed through a circuit

asbestos. A mineral fiber formed from a certain rock; a poor conductor of heat that can withstand high temperatures; used to insulate wires exposed to a high temperature

auxiliary. Extra, or something added to the main one

auxiliary bus. A second bus that may have a different voltage from the main bus and to which a few machines are connected

auxiliary circuit. Another circuit besides the main circuit; often a control circuit

B.t.u. British thermal unit

bakelite. A moulded insulating material

bayonet socket. A lamp socket that has two lengthwise slots in the sides; at the bottom the slots make a right-angle turn. The lamp base has two pins in it that slide in the slots in the socket. The lamp is held in the socket by being given a slight turn when the pins reach the bottom of the slots.

bell-ringing transformer. A small transformer which slips the voltage down from 110 volts to about 10 volts; used on a doorbell

box connector. An attachment used for fastening the ends of cable to a box

braided wire. A conductor composed of a number of small wires twisted or braided together

branch circuit. That part of the wiring system between the final set of fuses protecting it and the place where the lighting fixtures or drop cords are attached

British thermal unit. The amount of heat required to raise the temperature of one pound of water one degree F.

cabinet. Iron box containing fuse, cutouts, and switches

cable. A conductor composed of a number of wires twisted together

cable box. A box which protects the connections or splices joining cables of one circuit to another

cable clamp. A clamp used to fasten cables to their supports

cable grip. A clamp that grips the cable when it is being pulled into place

cable rack. A frame for supporting electric cables

capacity. Ability to hold or carry an electric charge; the unit of capacity is farad or microfarad

carrying capacity. The amount of current a wire can carry without overheating

cartridge fuse. A fuse enclosed in an insulating tube in order to confine the arc or vapor when the fuse blows

circuit. The path taken by an electrical current in flowing through a conductor from one terminal of the source of supply to the other

circuit breaker. A device used to open a circuit automatically

circular mil. The area of a circle one-thousandth of an inch in diameter; area in circular mils = diameter, in mils, squared (multiplied by itself)

closed circuit. A complete electric circuit through which current will flow when voltage is applied

conductivity. The ability of a substance to carry an electric current

conduit. A pipe or tube, made of metal or other material, in which electrical conductors or wires are placed

conduit box. An iron or steel box located between the ends of the conduit where the wires or cables are spliced

conduit bushing. A short threaded sleeve fastened to the end of the conduit inside the outlet box; inside of sleeve is rounded out on one end to prevent injury to the wires

conduit coupling. A short metal tube threaded on the inside and used to fasten two pieces of conduit end to end

conduit elbow. A short piece of conduit bent to an angle, usually to 45 or 90 degrees

conduit rigid. A mild steel tubing used to enclose electric wires

conduit wiring. Electric wires placed inside conduit

contactor. A device used to open and close an electrical circuit rapidly and often

continuous rating. The output at which a machine can operate continuously without overheating or exceeding a certain temperature

contractor. One who does a certain job for a sum of money agreed upon before the work is started

control switch. A small switch used to open and close a circuit which operates a motor or an electromagnet coil; this motor or electromagnet is used to operate or control some electric machine

controller. A device that governs or controls the action of electrical machines connected to it

copper. A metal used for electrical conductors because it has less resistance than any other metal except silver

cord. Two insulated flexible wires or cables twisted or held together with a covering of rubber, tape, or braid

corrosion. The rusting of iron; a similar action and deposit formed on other metals

counter-clockwise rotation. Turning in a direction opposite to the way of the hands of a clock usually turn

D.C. Direct current. Also used as an abbreviation for double contact

D.P. Double pole

D.P.S. Double pole snap switch

D.P.S.T. Double pole single throw

D.P.D.T. Double pole double throw

dead end. The end of a wire to which no electrical connection is made; the end used for supporting the wire; the part of a coil or winding that is not in use

dead wire. A wire in which there is no electric current or voltage

demand. Amount of electric current needed from a circuit or generator

dimmer. A resistance coil connected in series with a lamp to reduce the amount of current flowing through it, consequently dimming or reducing the light

direct current. Electric current flowing over a conductor in one direction only

disconnect. To remove an electrical device from a circuit, or to unfasten a wire, making part or all of the circuit inoperative

distribution box. Small metal box in a conduit installation, giving accessibility for connecting branch circuits

distribution. Division of current between the branches of an electrical circuit

distribution lines. The main feed line of a circuit to which branch circuits are connected

distribution panel. Insulated board from which connections are made between the main feed lines and branch lines

distribution system. The whole circuit and all of its branches which supply electricity to consumers

duct. A space in an underground conduit to hold a cable or conductor

duplex cable. Cable consisting of two wires insulated from each other and having a common insulation covering both

E. Symbol for volts

E.M.F. Electromotive force

earth. Synonym of "ground," meaning the grounded side of an electrical circuit or machine

efficiency. The ratio of the amount of power or work obtained from a machine and the amount of power used to operate it

electric circuit. Path through which an electric current flows

electric heater. Heater consisting of resistance wire which becomes hot as the current flows through it

electrical codes. Rules and regulations for the installation and operation of electrical devices and currents

electrocuted. Killed by electricity

electron. Electrical particle of negative polarity

extension. Length of cable or lamp cord fitted with a plug and a socket to extend a lamp or other electric device farther than the original point

F. Frequency

Fahrenheit. A thermometer scale so graduated that the freezing point of water is 32° and the boiling point is 212°

farad. Unit for measuring electrical capacity; the capacity of a capacitor which will give a pressure of one volt when a one-ampere current flows into it for one second

feeder. Line supplying all the branch circuits with the main supply of current

feeder box. Box into which the feeder is run for connection to a branch circuit

fish paper. Strong paper used for insulation

fish wire. Flat, narrow, flexible steel wire used to pull conductors through lengths of conduit

fixture. Device for holding electric lamps which is wired inside and is attached to the wall or ceiling

fixture wire. Insulated, stranded wire used for wiring fixtures

flexible cable. Cable consisting of insulated, stranded, or woven conductors

flexible conduit. Non-rigid conduit made of fabric or metal strip wound spirally

flexible cord. Insulated conductor consisting of stranded wire

flood lights. Battery of lamps of high brilliancy, equipped with reflectors to supply a strong light

flush receptacle. Type of lamp socket, the top of which is flush with the wall into which the socket is recessed

flush switch. Pushbutton or key switch, the top of which is flush with the wall into which it is recessed

four-way switch. Switch that controls the current in four conductors by making or breaking four separate contacts

four-wire, three-phase system. Distribution system having a three-phase star connection, one lead being taken from the end of each winding and the fourth from the point where they are all connected together

frequency. Number of cycles or vibrations per second

fuse. Safety device to prevent overloading a current; it consists of a short length of conducting metal which melts at a certain heat and thereby breaks the circuit

fuse block. Insulated block designed to hold fuses

fuse clip. Spring holder for a cartridge-type fuse

fuse link. An open fuse, or a length of fuse wire for refilling fuses

fuse plug. Fuse mounted in a screw plug, which is screwed in the fuse block like a lamp in a socket

gauge. Measuring device

Ground. See earth.

ground circuit. Part of an electric circuit in which the ground serves as a path for the current

ground clamp. Clamp on a pipe or other metal conductor connected to the ground for attaching a conductor of an electrical circuit

ground detector. Device used in a power station to indicate whether part of the circuit is accidentally grounded

ground indicator. Same as ground detector

ground plate. Metal plate buried in moist earth to make a good ground contact for an electrical circuit

ground return. Ground used as one conductor of an electrical circuit

ground wire. Conductor connecting an electrical device or circuit to the ground

grounded neutral wire. The neutral wire connected to the ground in a three-way distribution system

guy. A wire, rope, chain, or similar support for a structure such as a telephone pole, radio mast, etc.

h.p. Horsepower

heating unit. That part of a heating appliance through which the current passes and produces heat

high tension. High voltage

hot conductor. A conductor or wire which is carrying a current or voltage

humming. A noise caused by the rapid magnetizing and demagnetizing of the iron core of a transformer, motor, or generator

I. Symbol for amperes of current

illumination. The directing of light from its source to where it can be used to the best advantage

inside wiring. The wiring inside of a residence or building

insulate. To place insulation around conductors or conducting parts

insulating compound. An insulating wax which is melted and poured around electrical conductors in order to insulate them from other objects

insulating joint. A thread or coupling in which the two parts are insulated from each other

insulation resistance. The resistance offered by an insulating material to the flow of electric current through it

insulating varnish. A special prepared varnish which has good insulating property and is used to cover the coils and windings on electric machines and improve the insulation.

insulator. A device used to insulate electric conductors.

interior wiring. Wiring placed on the inside of buildings

I^2R loss. The power loss due to the current flowing through the conductor which has resistance; this loss is converted into heat

joint. The uniting of two conductors by means of solder

junction box. A box in a street distribution system where one main is connected to another main; also a box where a circuit is connected to a main

key switch. A switch for turning on and off electric circuits which are operated by means of a special key

key socket. A socket with a device that opens and closes the circuit, thus turning the lamp on or off

kilowatt. One thousand watts

knife switch. A switch having a thin blade that makes contact between two flat surfaces or short blades to complete the circuit

lamp. A device used to produce light

lamp base. The metal part of a lamp which makes contact with the socket

lamp bulb. An electric lamp

lamp circuit. A branch circuit supplying current to lamps only, and not to motors

lamp cord. Two flexible stranded insulated wires twisted together and used to carry the current from the outlet box to the lamp socket

lamp dimmer. An adjustable resistance connected to a lamp circuit in order to reduce the voltage and the brightness of the lamps

lamp socket. A receptacle into which the base of the lamp is inserted, which makes connection from the lamp to the circuit

leads. Short lengths of insulated wires that conduct current to and from a device

lighting fixture. An ornamental device that is fastened to the outlet box in the ceiling and which has sockets for the lamps

lighting transformer. A transformer used to supply a distribution circuit that does not have motors connected to it

lightning arrester. A device that allows the lightning to pass to the ground, thus protecting electrical machines

link fuse. A fuse that is not protected by an outside covering

live. A circuit carrying a current or having a voltage on it

load. The work required to be done by a machine. The current flowing through a circuit

load factor. The average power consumed divided by the maximum power in a given time

luminarre. An ornamental electric lighting fixture

luminosity. The relative quantity of light; brightness

main. The circuit from which all other smaller circuits are taken

main feeder. A feeder supplying power from the generating station to the main

maintenance. Repairing and keeping in working order

manual. Operated by hand

metal conduit. Iron or steel pipe in which electric wires and cables are installed

micro-. A prefix meaning one-millionth part

micro-ampere. One-millionth of an ampere; 1/1,000,000 or 0.000001 amperes

milli-. Prefix to a unit of measurement, denoting one-thousandth part of it

milli-ammeter. An instrument that reads the current in thousandths of an ampere

milli-ampere. 1/1,000 or 0.001 amperes; one-thousandth of an ampere

N.E.C. National Electric Code; often called "Underwriters Code"

negative. The point toward which current flows in an external electrical circuit; opposite of positive

neon. An inert gas used in electric lamps

network. A number of electrical circuits or distribution lines joined together

neutral. Not positive or negative, although it may act as positive to one circuit and negative to another

neutral conductor. A middle conductor of a three-wire direct-current or single-phase circuit

neutral terminal. A terminal which may be positive to one circuit and negative to another circuit

neutral wire. That wire in a three-wire distribution circuit which is positive to one circuit and negative to the other

nichrome. An alloy of nickel and chromium which forms a resistance wire that can be used at a high temperature

ohm. The unit used to express the resistance of a conductor to the flow of electric current through it

Ohm's law. A rule that gives the relation between current, voltage, and resistance of an electric circuit

open circuit. A break in a circuit; not having a complete path or circuit

outlet. A place where electrical wires are exposed so that one can be joined to the other

outlet box. An iron box placed at the end of conduit where electric wires are joined to one another and to the fixtures

overhead. Electric light wires carried outdoors on poles

overload. Carrying a greater load than the machine or device is designed to carry

overload capacity. The amount of load beyond a rated load that a machine will carry for a short time without dangerously overheating

P. Symbol for power

panel box. The box in which switches and fuses for branch circuits are located

parallel. Two lines which are equally distant at all points. Connecting machines or devices so that the current flows through each one separately from one line wire to another line wire

parallel circuit. A multiple circuit; a connection where the current divides and part flows through each device connected to it

peak load. The highest load on a system, or generator, occurring during a particular period of time

pilot lamp. A small lamp used on switchboards to indicate when a circuit switch or device has operated

plug. A screw thread device that screws into an electric light socket and completes the connection from the socket to the wires fastened to the plug

porcelain. A hard insulating material made from sand and clay which is moulded into shape and baked

positive. The point in a circuit from which the current flows; opposite of negative

power. The rate of doing work. In direct current circuits it is equal to $E \times I$. The electrical unit is the watt.

primary. That which is attached to a source of power, as distinguished from the secondary

raceways. Metal moulding or conduit that has a thinner wall than standard rigid conduit used in exposed wiring

rating. The capacity or limit of load of an electrical machine expressed in horsepower, watts, volts, amperes, etc.

receptacle. A device placed in an outlet box to which the wires in the conduit are fastened, enabling quick electrical connection to be made by pushing an attachment plug into it

receptacle plug. A device that enables quick electrical connection to be made between an appliance and a receptacle

return circuit. The path the current takes in going from the apparatus back to the generator

series circuit. A circuit in which the same current flows through all the devices

service connections. The wiring from the distributing mains to the building

service switch. The main switch which connects all the lamps or motors in a building to the service wires

service entrance. The place where the service wires are run into a building

service wires. The wires that connect the wiring in a building to the outside supply wires

sheath. The outside covering which protects a wire or cable from injury

short. A contraction for short circuit

short circuit. An accidental connection of low resistance joining two sides of a circuit, through which nearly all the current will flow

sine wave. The most perfect wave form; an alternating-current wave form

snap switch. A rotary switch where the contacts are operated quickly by a knob winding up a spring

socket. A receptacle or device into which a lamp bulb is placed

splice. The joining of the ends of two wires or cables together

step-down. Reducing from a higher to a lower value

step-up. Increasing, or changing from a low to a higher value

stranded wires. Wires or cables composed of a number of smaller wires twisted or braided together

surges. Oscillating high-voltage and current waves that travel over a transmission line after a disturbance

switch. A device for closing, opening, or changing the connections of a circuit

switch blade. The movable part of a switch

switch plate. A small plate placed on the plastered wall to cover a pushbutton or tumbler switch

sybmol. A letter, abbreviation, or sign that stands for a certain unit or thing

terminal lug. A lug soldered to the end of a cable so it can be bolted to another terminal

test lamp. An incandescent lamp bulb and socket connected in a circuit temporarily when making tests

thermocouple. Two different metals welded together and used for the purpose of producing thermo-electricity

three-way switch. A switch with three terminals by which a circuit can be completed through any one of two paths

three-wire circuit. A circuit using a neutral wire in which the voltage between outside wires is twice that between neutral and each side

time switch. A switch controlled by a clock that opens and closes a circuit at the desired time

timer. A device that opens the primary circuit of an induction coil at the right time to produce a spark to fire the charge in an internal combustion engine

transformer. A device used to change alternating current from one voltage to another. It consists of two electrical circuits joined together by a magnetic circuit formed in an iron core

underground cable. A cable insulated to withstand water and electrolysis and placed in underground conduit

Underwriters' Code. The National Electric Code

V. Symbol for volts or potential difference

volt. A unit of electrical pressure or electromotive force

voltage drop. The difference in pressure between two points in a circuit caused by the resistance opposing the flow of current

voltage loss. Voltage drop

voltage regulator. A device for keeping a constant voltage at a certain point

W. Symbol for watt

wall box. A metal box for switches, fuses, etc., placed in the wall

wall socket. An electric outlet placed in the wall so that conductors can be connected to it by means of a plug

watt. The unit of electric power

watt-hour. The use of a watt of power for an hour

watt-hour meter. An instrument that records the power used in watt-hours

watt meter. An instrument used to indicate the power being used in a circuit

wire. A slender rod of drawn metal

wire gauge. A method of measuring the diameter of different wires

wiring connector. A device for joining wire to another

wiring symbols. Small signs placed on a wiring diagram to indicate different devices and connections

INDEX